Praise for Thomas Gold. . .

"Gold is one of America's most iconoclastic scientists."
—Stephen Jay Gould

"Thomas Gold is one of the world's most original minds."
—*The Times*, London

"Thomas Gold might have grown tired of tilting at windmills long ago had he not destroyed so many."
—*USA Today*

"What if someone told you that [the oil crisis] was all wrong and that the hydrocarbons that make up petroleum are constantly refilling reservoirs. Interested? Well, you should read this book. . . . Gold presents his evidence skillfully. You may not agree with him, but you have to appreciate his fresh and comprehensive approach to these major areas of Earth science. . . . [This book] demonstrates that scientific debate is alive and well. Science is hypothesis-led and thrives on controversy—and few people are more controversial than Thomas Gold."
—*Nature*

". . . Thomas Gold, a respected astronomer and professor emeritus at Cornell University in Ithaca, N.Y., has held for years that oil is actually a renewable, primordial syrup continually manufactured by the Earth under ultrahot conditions and pressures."
—*The Wall Street Journal*

"Most scientists think the oil we drill for comes from decomposed prehistoric plants. Gold believes it has been there since the Earth's formation, that it supports its own ecosystem far underground and that life there preceded life on the Earth's surface. . . . If Gold is right, the planet's oil reserves are far larger than policymakers expect, and earthquake prediction procedures require a shakeup; moreover, astronomers hoping for extraterrestrial contacts might want to stop seeking life *on* other planets and inquire about life *in* them."
—*Publishers Weekly*

"Gold's theories are always original, always important, and usually right. It is my belief, based on fifty years of observation of Gold as a friend and colleague, that *The Deep Hot Biosphere* is all of the above: original, important, controversial, and right."
—Freeman Dyson

"Whatever the status of the upwelling gas theory, many of Gold's ideas deserve to be taken seriously. . . . The existence of [a deep hot biosphere] could prove to be one of the monumental discoveries of our age. This book serves to set the record straight."
—*Physics World*

"My knowledge and experience of natural gas, gained from drilling and operating many of the world's deepest and highest pressure natural gas wells, lends more credence to your ideas than the conventional theories of the biological/thermogenic origin of natural gas. Your theory explains best what we actually encountered in deep drilling operations."
—Robert A. Hefner III, The GHK Companies, Oklahoma City, Oklahoma;
From a letter to the author

"Within the scientific community, Gold has a reputation as a brilliantly clever renegade, having put forward radical theories in fields ranging from cosmology to physiology."
—*The Sunday Telegraph*, London

"In *The Deep Hot Biosphere*, [Gold] reveals evidence supporting a subterranean biosphere and speculates on how energy may be produced in a region void of photosynthesis. He speculates on the ramifications his concepts could have in predicting earthquakes, deciphering Earth's origins, and finding extraterrestrial life."
—*Science News*

"Gold's theory, as explained in *The Deep Hot Biosphere*, offers new and radical ideas to our incomplete notions of what causes earthquakes and where we would look for life in outer space: not on planets, but in them."
—*Ithaca Times*

"[*The Deep Hot Biosphere*] now seems to be supported by a growing body of evidence."
—*Journal of Petroleum Technology*

"Gold knows experts are pooh-poohing his belief. It happens to Gold consistently. He has developed a reputation as someone who takes on a long-held assumption, advances a new idea and gets rewarded when time—a decade or two—proves him right."
—*The Juneau Empire*

"Thomas Gold has questioned the very foundations of the entrenched conventional models. . . . [*The Deep Hot Biosphere*] is evidently one of the most controversial of all books published in recent history. It is bound to cause much debate, and, if found correct, is likely to revolutionize the face of science."
—*Current Science*

"[Thomas Gold] is one of the few who, despite the attacks of mediocrities, is courageous enough to think in a scientifically unconventional way. . . . [His] courage and original ideas are rays of hope on the horizon of science."
—Prof. Dr. Alfred Barth, The European Academy of Sciences and Arts, Paris

The Deep Hot Biosphere
The Myth of Fossil Fuels

Thomas Gold

With a Foreword by Freeman Dyson

COPERNICUS BOOKS

AN IMPRINT OF SPRINGER-VERLAG

First softcover printing, 2001

Published in the United States by Copernicus Books,
an imprint of Springer-Verlag New York, Inc.
A member of BertelsmannSpringer Science+Business Media GmbH

    Copernicus Books
    37 East 7th Street
    New York, NY 10003
    www.copernicusbooks.com

Library of Congress Cataloging-in-Publication Data
Gold, Thomas.
    The deep hot biosphere: the myth of fossil fuels / Thomas Gold ; foreword
by Freeman Dyson.
        p. cm.
    Includes bibliographical references and index.
    ISBN 0-387-95253-5 (softcover ; alk. paper)
        1. Deep-earth gas theory.   2. Petroleum—Geology.
    3. Hydrocarbons.   4. Life—Origin.   5. Extreme environment
    microbiology.   I. Title.
TN870.5.G66   1998
576.8'3—dc21   98-42598

9 8 7 6 5 4 3 2 1

ISBN 0-387-95253-5   SPIN 10795990

Foreword

by Freeman Dyson

———————————————————

T he first time I met Tommy Gold was in 1946, when I served as a guinea pig in an experiment that he was doing on the capabilities of the human ear. Humans have a remarkable ability to discriminate the pitch of musical sounds. We can easily tell the difference when the frequency of a pure tone wobbles by as little as 1 percent. How do we do it? This was the question that Gold was determined to answer. There were two possible answers. Either the inner ear contains a set of finely tuned resonators that vibrate in response to incident sounds, or the ear does not resonate but merely translates the incident sounds directly into neural signals that are then analyzed into pure tones by some unknown neural process inside our brains. In 1946, experts in the anatomy and physiology of the ear believed that the second answer must be correct: that the discrimination of pitch happens in our brains, not in our ears. They rejected the first answer because they knew that the inner ear is a small cavity filled with flabby flesh and water. They could not imagine the flabby little membranes in the ear resonating like the strings of a harp or a piano.

Gold designed his experiment to prove the experts wrong. The experiment was simple, elegant, and original. During World War II he had been working for the Royal Navy on radio communications and radar. He built his apparatus out of war surplus Navy electronics and headphones. He fed into the headphones a signal consisting of short pulses of a pure tone, separated by intervals of silence. The silent intervals were at least ten times longer than the period of the pure tone. The pulses were all the same shape, but they had phases that could be reversed independently. To reverse the phase of a pulse means to reverse the movement of the speaker in the headphone. The speaker in a reversed pulse is pushing the air outward when the speaker in an unreversed pulse is pulling the air inward. Sometimes Gold gave all the pulses the same phase, and sometimes he alternated the phases so that the even pulses had one phase and the odd pulses had the opposite phase. All I had to do was sit with the headphones on my ears and listen while Gold fed in signals with either constant or alternating phases. Then I had to tell him, from the sound, whether the phase was constant or alternating.

When the silent interval between pulses was ten times the period of the pure tone, it was easy to tell the difference. I heard a noise like a mosquito, a hum and a buzz sounding together, and the quality of the hum changed noticeably when the phases were changed from constant to alternating. We repeated the trials with longer silent intervals. I could still detect the difference, even when the silent interval was as long as thirty periods. I was not the only guinea pig. Several other friends of Gold listened to the signals and reported similar results. The experiment showed that the human ear can remember the phase of a signal, after the signal stops, for thirty times the period of the signal. To be able to remember phase, the ear must contain finely tuned resonators that continue to vibrate during the intervals of silence. The result of the experiment proved that pitch discrimination is done mainly in the ear, not in the brain.

Besides having experimental proof that the ear can resonate, Gold also had a theory to explain how a finely tuned resonator can be built out of flabby and dissipative materials. His theory was that the inner ear contains an electrical feedback system. The mechanical resonators are coupled to electrically powered sensors and drivers, so that the combined electromechanical system works like a finely tuned amplifier. The positive feedback provided by the electrical components counteracts the damping produced by the flabbiness of the mechanical components. Gold's experience as an electrical engineer made this theory seem plausible to him, although he could not identify the anatomical structures in the ear that functioned as sensors and drivers. In 1948 he published two papers, one reporting the results of the experiment and the other describing the theory.

Having myself participated in the experiment and having listened to Gold explaining the theory, I never had any doubt that he was right. But the professional auditory physiologists were equally sure that he was wrong. They found the theory implausible and the experiment unconvincing. They regarded Gold as an ignorant outsider intruding into a field where he had no training and no credentials. For years his work on hearing was ignored, and he moved on to other things.

Thirty years later, a new generation of auditory physiologists began to explore the ear with far more sophisticated tools. They discovered that everything Gold had said in 1948 was true. The electrical sensors and drivers in the inner ear were identified. They are two different kinds of hair cells, and they function in the way Gold said they should. The community of physiologists finally recognized the importance of his work, forty years after it was published.

Gold's study of the mechanism of hearing is typical of the way he has worked throughout his life. About once every five years, he invades a new field of research and proposes an outrageous theory that arouses intense opposition from the professional experts in the field. He then works very hard to prove the experts wrong. He does not always succeed. Sometimes it turns out that the experts are right and he is wrong. He is not afraid of being wrong. He was famously wrong (or so it is widely believed) when he promoted the theory of a steady-state universe in which matter is continuously created to keep the density constant as the universe expands. He may have been wrong when he cautioned that the moon may present a dangerous surface, being covered by a fine, loose dust. It proved indeed to be so covered, but fortunately no hazards were encountered by the astronauts. When he is proved wrong, he concedes with good humor. Science is no fun, he says, if you are never wrong. His wrong ideas are insignificant compared with his far more important right ideas. Among his important right ideas was the theory that pulsars, the regularly pulsing celestial radio-sources discovered by radio-astronomers in 1967, are rotating neutron stars. Unlike most of his right ideas, his theory of pulsars was accepted almost immediately by the experts.

Another of Gold's right ideas was rejected by the experts even longer than his theory of hearing. This was his theory of the 90-degree flip of the axis of rotation of the earth. In 1955, he published a revolutionary paper entitled "Instability of the Earth's Axis of Rotation." He proposed that the earth's axis might occasionally flip over through an angle of 90 degrees within a time on the order of a million years, so that the old north and south poles would move to the equator, and two points of the old equator would move to the poles. The flip would be triggered by movements of mass that would cause the old axis of rotation to become unstable and the new axis of rotation to become stable. For example, a large accumulation of ice at the old north and south poles might cause such an exchange of stability. Gold's paper was ignored by the experts for forty years. The experts at that time were focusing their attention narrowly on the phenomenon of continental drift and the theory of plate tectonics. Gold's theory had nothing to do with continental drift or plate tectonics, so it was of no interest to them. The flip predicted by Gold would occur much more rapidly than continental drift, and it would not change the positions of continents relative to one another. The flip would change the positions of continents only relative to the axis of rotation.

In 1997, Joseph Kirschvink, an expert on rock magnetism at the California Institute of Technology, published a paper presenting evidence that a 90-degree flip of the rotation axis actually occurred during a geologically short time in the early Cambrian era. This discovery is of great importance for the history of life, because the time of the flip appears to coincide with the time of the "Cambrian Explosion," the brief period when all the major varieties of higher organisms suddenly appear in the fossil record. It is possible that the flip of the rotation axis caused profound environmental changes in the oceans and triggered the rapid evolution of new life forms. Kirschvink gives Gold credit for suggesting the theory that makes sense of his observations. If the theory had not been ignored for forty years, the evidence that confirms it might have been collected sooner.

Gold's most controversial idea is the non-biological origin of natural gas and oil. He maintains that natural gas and oil come from reservoirs deep in the earth and are relics of the material out of which the earth condensed. The biological molecules found in oil show that the oil is contaminated by living creatures, not that the oil was produced by living creatures. This theory, like his theories of hearing and of polar flip, contradicts the entrenched dogma of the experts. Once again, Gold is regarded as an intruder ignorant of the field he is invading. In fact, Gold *is* an intruder, but he is not ignorant. He knows the details of the geology and chemistry of natural gas and oil. His arguments supporting his theory are based on a wealth of factual information. Perhaps it will once again take us forty years to decide whether the theory is right. Whether the theory of non-biological origin is ultimately found to be right or wrong, collecting evidence to test it will add greatly to our knowledge of the earth and its history.

Finally, the most recent of Gold's revolutionary proposals, the theory of the deep hot biosphere, is the subject of this book. The theory says that the entire crust of the earth, down to a depth of several miles, is populated with living creatures. The creatures that we see living on the surface are only a small part of the biosphere. The greater and more ancient part of the biosphere is deep and hot. The theory is supported by a considerable mass of evidence. I do not need to summarize this evidence here, because it is clearly presented in the pages that follow. I prefer to let Gold speak for himself. The purpose of my remarks is only to explain how the theory of the deep hot biosphere fits into the general pattern of Gold's life and work.

Gold's theories are always original, always important, usually controversial—and usually right. It is my belief, based on fifty years of observation of Gold as a friend and colleague, that the deep hot biosphere is all of the above: original, important, controversial—and right.

Preface

In June 1997 I was asked by NASA to give the annual lecture at the Goddard Space Flight Center in Maryland. My contribution to the deep hot biosphere theory and its implications for extraterrestrial life had won me the invitation. I was flattered, of course, but at the same time chagrined by the topic I was asked to address: life in extreme environments. I had little interest in talking about the surface biosphere on earth, and yet, if I were to take the topic literally, this is precisely what I was being asked to do. The life in extreme environments is our own surface life.

If there is one idea that I hope you will retain long after you finish reading this book, it is this: It is we who live in the extreme environments. And if there is one desire I hope to stimulate in you, it is a curiosity to learn more about the first and most truly *terrestrial* beings—all of whom live far beneath our feet, in what I have come to call the deep hot biosphere.

Alas, I can only begin to satisfy this curiosity here, for at this moment in our biological and cosmic understanding, there are still more questions than answers. But that is exactly what makes investigating the deep hot biosphere so exciting.

Thomas Gold
Ithaca, New York
December 1998

Contents

Chapter 1 Our Garden of Eden

———————————————

No scientific subject holds more surprises for us than biology. Foremost is the surprise that life exists at all. How could life have started? Did one extraordinary chance occurrence in the universe assemble the first primitive living organism, and did everything else follow from that?

What chemical and physical circumstances were needed for such an unlikely event to occur? Did our earth offer the only nurturing conditions? Or (in what has come to be known as the "panspermia" hypothesis) did life arise somewhere else, spreading through astronomical space to take root in any fertile spot it encountered? Or is life not unlikely after all? Perhaps life is an inevitable consequence of physical laws and is arising spontaneously in millions of places.

Whatever the answers to these questions, we do know that life on the surface of the earth spans a huge variety of forms. These forms range from microbes to whales, giant fungi, and enormous trees. They include unfathomable numbers of insects. If we add to our reckoning the life forms that have died out, then the diversity expands to include dinosaurs, trilobites, and vastly more.

All this living variety has much in common. The construction of all known organisms involves complex forms of protein molecules. Those, in turn, are built up from a set of building blocks called amino acids, common to all known forms of life. The chemical configuration of some of these amino acids *could* occur in two forms, one of which is the mirror image of the other. Yet we find that all the huge variety of life uses only one kind of each such pair of molecules. There thus appears to be a strong connecting thread running through all the life forms we know.

No less important than the common constituents of life are the common conditions under which all known life forms can develop and survive. These conditions include a requirement for water in the liquid state, a limited range of temperature, and sources of energy that are delivered in (or can be converted into) chemical form. We tend to assume that these conditions are best—and perhaps ideally—provided on the surface of our own planet. And we conclude, sadly perhaps, that these conditions are almost certainly not present anywhere else in the solar system. But are these assumptions valid?

The Narrow Window for Surface Life

The universe is a harsh and severe place, a realm of extremes. Most of the universe is virtually empty and very cold—to be precise, 2.7 Kelvin or −270.5° Celsius, which is just 2.7°C above absolute zero. This vast cold is punctuated by points of intense heat and light—the stars—whose surface temperatures reach millions of degrees.

Stars do not maintain their brilliance forever, and it is from them that the constituents of life come. Stars that have three or more times the mass of the sun will expire in a frenzy of violence, a supernova explosion that may briefly flare with the brightness of a hundred billion stars. The explosion scatters the stellar materials into space, making the cold clouds out of which new stars form. The different atomic nuclei created in the core of the star and during its explosion supply materials from which planets can form. The same stellar materials provide the elements from which we and all other living creatures known to us are constructed.

Life is thus built up from a variety of atoms forged in nuclear furnaces deep inside giant stars. More precisely, life is constructed from molecules, clumpings of atoms that are in close enough contact and cool enough for a weak attractive force to hold them together. The interiors of stars are suitable for element formation, but their heat is too intense for the formation of complex molecules.

Most places in the universe do not allow the chemical action that is conducive to life. The stars are too hot, and most other places are so cold that substances are in the form of a solid or a very low-density gas, whose chemical activity is exceedingly slow. But we do see some regions in the cosmos in which many different types of molecules have been built up. These are the large gas clouds in interstellar spaces, warmed by stars that are in or near them. Radio techniques have made it possible to identify many different molecules there. Water is one common component of the gas, as are hydrocarbons—combinations of hydrogen and carbon. It is from the materials of such clouds that our and other star systems are believed to have formed.

For life forms to arise and to persist, molecules must be awash in a liquid or a gas, so that gentle contacts among molecules can build up other molecules and generate a brew of the kind of complexity we find in biological materials. In all of the expressions of life known to us, this mobility is provided by liquid water. Given the ferocious and unfriendly conditions of the universe—with points of intense heat and vast expanses of severe cold—one would think it rare indeed for any place to hold surface temperatures in the range that would render water a liquid. Surface temperatures depend not only on the solar irradiation intercepted by the planet, and thus on its distance from the sun and on the sun's size and surface temperature, but also on the mass and composition of the planet's atmosphere.

It is the mass and composition of the atmosphere that crucially determines atmospheric pressure. Without a gas pressure, there is no such thing as liquid water. In the absence of substantial atmosphere, water is either a solid or a vapor. All in all, a planet that offers liquid water on its surface is a rare occurrence. Rarer still would be the subset of such places that have given rise to the intricate designs that we call "life."

Could there be, in this fierce universe, locations where perhaps a little brook runs down a hillside, with trees gently swaying in the wind, and with creatures sitting by the side, enjoying the view? It seems a far-fetched fantasy in this forbidding universe. And yet we know one such place: our little earth.

How was our planet able to bring forth the enormous abundance of surface life that we see around us? None of the other planets and none of their moons have anything comparable. Indeed, because the surfaces of all other bodies in our planetary system offer essentially no possibility for the existence of liquid water, it is very unlikely that surface life exists anywhere in our solar system other than on the earth. There may be only one Garden of Eden here for large life forms such as ourselves. But living beings small enough to populate tiny pore spaces may well exist *within* several—and perhaps many—other planetary bodies.

Chemical Energy for Subsurface Life

The sun provides two distinct actions. First, it is the source of heat that puts the surface temperature of the earth into a range suitable for the complex chemical reactions of molecules, and thus for life. But ambient heat cannot be a source of energy, and the warmth of our surface surroundings could not constitute an energy source for surface life. Only a heat *flow* from a hotter body to a cooler one can be converted into other forms of energy. We have such an energy flow from the hot surface of the sun to the cooler earth—the second action that the sun provides—and energy is taken from this *flow* and converted into chemical energy in the process of photosynthesis.

Photosynthesis is performed today largely by plants and algae, using sunlight to dissociate water molecules (H_2O) and atmospheric carbon dioxide (CO_2), then reconfiguring the atoms to yield carbohydrates such as $C_6H_{12}O_6$, which can than be oxidized ("burned") as needed, back into H_2O and CO_2, to yield metabolic energy. This process then serves as the principal energy source for all surface life. A planetary surface that does not possess photosynthetic life would be hostile to any of the surface life forms we know. Below the surface the temperature may be similar to that at the surface; but over small dimensions—like the size of living forms there—only quite insignificant energy *flow* occurs. Therefore, no energy source can exist beneath the earth's surface.

When we consider life's beginning, however, we realize that a puzzle lurks in this account of energy transformation. Photosynthesis is an exceedingly complex process. The microorganisms that developed it must have already possessed intricate chemical processing systems before they acquired this more advanced ability. The energy source that these initial microorganisms drew on must have been chemical to begin with. The chemical energy available before the advent of photosynthesis could not have been created by solar energy or by life. It must have been a free gift of the cosmos.

Where exactly did such chemical energy come from? I propose that the original source of energy for earthly life was derived not from photosynthesis but from the oxidation of hydrocarbons that were already present, just as they are also present on many other planetary bodies and in the original materials that formed the solar system. Spanning the range from the light gas methane to the heaviest petroleum, hydrocarbons are present in the earth today in large amounts and to great depths—I believe much larger and deeper than is typically estimated. This view of the genesis of hydrocarbons I have called the *deep-earth gas theory.*[1]

I think we have good evidence now that a very significant realm of life has existed, and still exists, well below the surface biosphere that is home to humans. This subsurface realm and its inhabitants constitute what I call the *deep hot biosphere*[2]—deep because it may extend down to a depth of ten kilometers or more below the surface of the earth, and hot because, as a result of the natural temperature gradient of the earth, temperatures in much of that realm approach and even exceed 100°C.

The conventional notion is that hydrocarbons present within the earth's upper crust are derived strictly from plant and animal debris transformed by geological processes—and thus that hydrocarbons could not possibly have played a role in the *origin* of life. But we shall have reason to question this, along with many other assumptions. And as we shall see in Chapter 2, an abundance of new discoveries have confirmed life's presence in this crustal realm and under conditions seldom before thought tolerable to any form of life.

Chemical energy is released in chemical reactions. The substances we call fuels in our surface realm are really only one component of the energy-producing reactions. The other component, oxygen, is so abundant around us that we tend to forget about it. Hydrocarbons, hydrogen, and carbon are fuels for us only because the other component needed for the reaction that produces energy is readily available from the vast store of oxygen present in our atmosphere and dissolved in seawater as O_2. This oxygen is largely, but not entirely, created as a residue substance in the process of photosynthesis. It, rather than the petroleum or the coal, represents the *fossil* fuel left over from bygone vegetation.

Before photosynthesis was devised by life—and even now at depths to which atmospheric oxygen cannot penetrate—any hydrocarbon-using life must have depended on other sources of oxygen. Oxygen is the second most abundant element (after silicon) in the crust of the earth. The rocks therefore have plenty of oxygen in them, but most of it is too tightly bound to be useful. Clearly, sources of oxygen that require more energy to free the oxygen from its attachment in the rocks than the energy gained by oxidizing hydrocarbons with it cannot provide microbes with an energy supply.

Subsurface life must therefore depend on sources of oxygen in which these vital atoms are only weakly bound with other elements. The largest sources of weakly bound oxygen in the earth's crust are certain kinds of iron oxides and sulfates (oxidized sulfur compounds). When oxygen is extracted from iron oxides such as ferric iron, that process leaves behind iron in a lower oxidation state in which it is magnetic; examples include the minerals magnetite and greigite. When oxygen is taken from sulfates, what is left behind may be pure sulfur or sulfides such as hydrogen sulfide and iron sulfide. The existence of such by-products of metabolic activity in the subsurface realm helps us identify the biochemical processes that have occurred. These by-products also provide a sense of the scale and reach of the deep hot biosphere.

It is crucial to the theory of subsurface life that the ultimate source of up-welling hydrocarbons resides very much deeper than the lowermost reach of subsurface life. The deep hot biosphere may be deep, but it must not be excessively deep. Why is this so? The exponential growth rates of microbes (as of all forms of life) mean that wherever life resides, the source of energy that supports it must arrive in a *metered* flow. If the earliest forms of subsurface life had not been checked by limits on their food supply, the increase in their numbers would have very rapidly consumed the entire lot in an instant of geological time, allowing no gradual evolution to take place.

Hence energy that can be used by life must be available, but it must not be available all at once. The metered energy flow for the surface biosphere is provided by a sun that takes billions of years to consume its own finite stores of fuels. The chemical energy (such as sugars) forged by photosynthesizing life forms here on the earth is thus created through time in a metered way and only in areas that have liquid water—not in the driest deserts or in the icefields of polar or high mountain regions. The transformation from solar to chemical energy now takes place at a rate sufficient to feed all the surface life we see. But no matter how greedy life may be, organisms simply cannot make the sun radiate energy any faster. It is energy that supports life, but only a metered flow of energy sustains life over a long period of time.

Understanding the importance to life of a metered supply of energy is crucial to delimiting the possibilities for life's origins. The often-discussed warm little pond that contained nutrients forged with great difficulty by surface processes is not a candidate environment, in my opinion, for the transition from non-life to life. Such an environment would yield a limited amount of chemical supplies and energy, not a long-term and continuous metered supply. What is needed, rather, is an environment that can supply chemical energy in a metered flow over tens or hundreds of millions of years, during which time incomprehensibly large numbers of molecular experiments might take place.

A Preview of This Book

In the remaining chapters, I shall set forth the theory that a fully functioning and robust biosphere, feeding on hydrocarbons, exists at depth within the earth and that a primordial source of hydrocarbons lies even deeper. I will argue that photosynthesis developed in off-shoots of subterranean life that had progressed toward the surface and then evolved a way to use photons to supply even more chemical energy. When surface conditions became favorable to life (with regard to temperature, the presence of liquid water, the filtering of harsh components of solar radiation, and the termination of devastating asteroid impacts), a huge amount of surface life was able to spring up.

In retrospect, it is not hard to understand why the scientific community has typically sought only *surface* life in the heavens. Scientists have been hindered by a sort of "surface chauvinism." And because earth scientists did not recognize the presence of chemical energy beneath their feet, astronomers and planetary scientists could not build a subsurface component into their quests for extraterrestrial life. Unfortunately, this misunderstanding lingers. The idea that hydrocarbons on earth are the chemical remains of surface life that has long been buried and pressure-cooked into petroleum and natural gas has been exceedingly difficult to unseat. I have been trying to do so since 1977, and I discovered along the way that some pioneering Russian scientists were my forebears.[3] The reason for this continuing confusion in understanding how hydrocarbons came into being is a story in itself; I shall take it up in Chapter 3.

As long as Western scientists continue to assume a biological origin for all terrestrial hydrocarbons, the major sources of the earth's chemical energy will not be recognized. And as long as this substantial food supply goes unrecognized, the prospect that a large subterranean biosphere may indeed exist, and exist down to great depth, will likewise fail to attract scientific attention. Thus the particular importance of Chapter 3, in which I will examine the considerations that favor the deep-earth gas theory.

Surface evidence for that theory follows in Chapter 4. Most important, I introduce a set of observations that cannot be explained at all by a sedimentary origin of hydrocarbons—the strong association of hydrocarbons with a gas that can have no chemical interactions either with plant materials or with hydrocarbons: the inert element helium. How can petroleum have gathered up clearly biological molecules but also an inert gas that is normally sparsely distributed in the rocks? I call this association the "petroleum paradox." Its resolution (in Chapter 5) suggests that multitudes of microbial life must exist in the pore spaces of the rocks. In my view, hydrocarbons are not biology reworked by geology (as the traditional view would hold) but rather geology reworked by biology. In other words, hydrocarbons are primordial, but as they upwell into earth's outer crust, microbial life invades.

Chapter 6 presents the striking results of a large-scale drilling project that I initiated in Sweden to test the deep-earth gas theory and also to look for deep microbial life. In Chapters 7 and 8, I undertake to show how the deep-earth gas theory can account for concentrated deposits of certain metal ores in the crust and also for important features of earthquakes.

In Chapters 9 and 10 I use the deep-earth gas and deep hot biosphere theories to offer new speculations on what are perhaps the two most profound mysteries of the biological sciences: the origin of earth life and the prospects for extraterrestrial life. As background, I begin with a comparison of the two biospheres. In what major ways might the surface biosphere and the deep biosphere differ, beyond the simple fact that one draws on chemical energy and the other on solar? I then revisit the question of life's origin, explaining why I believe that surface life is the descendant of an original form of life that began at depth, rather than the other way around.

If this sequence from depth to surface best explains the origin and expansion of terrestrial life, then subsurface life on many other planetary bodies would seem very probable. There are many bodies in the solar system whose internal conditions are thought to be similar to those of our earth but whose surfaces do not offer the extraordinary advantages for life that ours has. It would be unlikely indeed for subsurface life to develop just in the one unique body that could support surface life as well. This reasoning led me in 1992 to make the tentative prediction that our own solar system harbors not one but *ten* deep hot biospheres.[4]

We surface creatures may well be alone in the solar system, but the denizens of the terrestrial deep seem likely to have many—possibly independently evolved—peers. Only when we recognize the existence of a thriving subterranean biosphere within our own planet will we learn the right techniques to begin the search for extraterrestrial life in other planetary bodies. Some such techniques and further suggestions for future research will be presented in Chapter 10.

Our journey will begin in the next chapter with a look at the borderland regions between the two biospheres. Along hydrothermal vents and petroleum seeps of the ocean, and in hot springs and methane-rich caves on land, we encounter some extraordinary ambassadors from the deep hot biosphere. Here we can also begin to comprehend why deep may, in fact, be desirable for life.

Chapter 2 Life at the Borders

———————————

I n February and March of 1977, the small deep-sea-diving submarine *Alvin* descended to a depth of 2.6 kilometers along the East Pacific Rise. This region, northeast of the Galapagos Islands, was known to be a center of sea floor spreading. A research ship had drawn a camera over the area the previous year, confirming the existence of a series of cracks in the ocean floor that appeared to be volcanically active. But the occupants of *Alvin* saw much more.

Far below the deepest possibility for photosynthetic life, *Alvin*'s searchlight revealed a patch of ocean bottom teeming with life, in sharp contrast with the surrounding barrens. This patch was covered with dense communities of sea animals—some exceptionally large for their kind. Anchored to the rocks, these creatures thrived in the rich borderland where hot fluids from the earth met the marine cold. New to science were species of lemon-yellow mussels and white-shelled clams that approached a third of a meter in length. Most striking of all were the tube worms, which lurk inside vertical white stalks of their own making, bright red gills protruding from the top. Like the tube worms of shallow waters, these denizens of the deep live clustered together in communities, with tubes oriented outward resembling bristles on a brush. But unlike their more familiar kin, the tube worms of the deep are giants, reaching lengths in excess of two meters.

Further investigations soon revealed that this strange and isolated community of life was by no means unique. Populations of the same organisms were discovered at other points along that ocean rift, at hydrothermally active vents elsewhere in the Pacific, and in the Atlantic and Indian Oceans too. This was clearly a global phenomenon. These unsuspected oases represent an entirely new habitat for life. Where did these creatures come from? What sources of energy and nutrients could support such astonishing fecundity and in such a patchwork distribution?

Through the windows of *Alvin,* the 1977 discovery crew witnessed not only strange life forms but also streams of milky fluids and black "smoke" emerging from vents in the sea floor. These streams of hydrothermal fluids, heated and enriched in gases and minerals, are now known to be the sources of chemical energy at the base of the vent community's food chain. Two decades later, however, we have only begun to understand how it all works.

Because we are surface creatures, we readily adopt the outlook that surface life is the only possible kind. We marvel at the exotic life along the deep-ocean vents. We assume, of course, that the vents were originally colonized by emigrants from a surface ecosystem—pioneers in evolving the adaptations necessary to subsist on energy drawn from chemical sources rather than bundled in photons, the units of energy in which light is delivered. This top-down scenario is reasonable for the large animals. Tube worms and clams surely did migrate down from shallow waters. But no animal of any kind can serve as the base of a food chain. All animals depend on chemical energy stored in the bodies of organisms they consume. Something, therefore, must have already been growing around the ocean vents when the worms and clams arrived.

In my view, the base of the food chain in the deep ocean vents is more likely to have emerged from below than to have descended from above. The microbes (bacteria and archaea) that today support the whole complex enterprise are offspring of microbial communities that lived and still live within the earth's crust. Whereas the large life forms can exist only where there is considerable space for them, the microbial life that feeds them occurs in units small enough to inhabit minute cracks in the rocks of the sea floor and elsewhere throughout the earth's upper crust. The total volume of rock that is accessible to such microbes is enormous; as we shall see in Chapter 5, the microbial content of the earth's upper crust may well exceed in mass and volume all surface life. Indeed, microbes from the realm that I call the deep hot biosphere probably invaded this borderland between the two worlds—between the deep biosphere and the surface biosphere—long before photosynthesis evolved on the surface. In fact, the chemical differences between the two worlds may have been slight prior to the advent of photosynthesis, because it was photosynthesis that transformed the earth's surface into a zone pervaded by free oxygen—molecules of O_2.

Energy Deep in the Earth

Photosynthesis is an exceedingly complex process for turning the energy of light into chemical energy. But why does the route that energy takes have to include chemical forms? Why cannot the sunlight be made to drive directly all the processes that the organism requires? There are some compelling reasons. First, the energy required to run cellular metabolisms must be available in increments no more than a tenth as powerful as that supplied by even a single solar photon. Expecting a cell to use a photon directly to synthesize a sugar would be more ludicrous than expecting a baseball player to field bullets from a machine gun. Rather, life has devised an extremely sophisticated apparatus to perform the initial task of catching the bullets.

Second, a photon has no patience. Make use of it now or lose it forever. Sunlight cannot be captured in a jar and stored on a shelf. But its energy can be used to set up molecules such as sugars, that will deliver energy on combining with atmospheric oxygen. Our breathing demonstrates this: we take in such "reduced" (unoxidized) carbon compounds in our food and we inhale oxygen and exhale carbon dioxide. This describes the overall metabolic activity, but in fact there are various stages in between, all dependent on the energy provided by the oxidation of the reduced carbon compounds we eat, eventually to CO_2. Sugars or other intermediate molecules can be stored on the cellular shelf, and the rate of "combustion" can be controlled. Chemical energy thus carries the advantage of availability, offering an adjusted amount where and when it is needed.

Because photosynthesis is such a complex process, and because the energy derived from photons must be converted into chemical energy before the cell can make use of it, researchers who probe the possible origins of earthly life have become convinced that the first living cells tapped not sunlight but chemical energy present in the environment. Where this chemical energy came from and what it consisted of remain hotly debated issues, but the widespread assumption is that either this primordial energy source has long since been used up or the conditions that produced it billions of years ago no longer prevail. I shall return to this question in later chapters. For now it is important to remember only that it would be far more difficult to design a living cell that could construct chemical energy from photons than it would be to design a living cell that scavenged chemical energy from its surroundings.

The cells that perform this complex function of photosynthesis must have access to liquid water, as already noted, and they must have access to carbon and nitrogen for the fabrication of proteins, the principal building blocks for their chemical machinery. The solar energy is used to "reduce" (unoxidize) compounds that will serve to provide energy as they are later oxidized again. Oxygen must therefore also be available, as must catalysts (enzymes) that initiate and control the reaction rates and thereby the power output.

Life as we know it depends fundamentally on the presence of carbon; earth life is sometimes referred to as "carbon-based life," to distinguish it from the theoretically possible (but unknown) "silicon-based life." Carbon atoms constitute the skeletal structure of all proteins and of all genetic materials of all the life forms we know. In the surface biosphere, carbon is provided by carbon dioxide, which is present in small proportion in the atmosphere. Each of the several varieties of photosynthesis that life has evolved begins with carbon dioxide, from which the complex molecules of life are then forged. In the most common form of photosynthesis, energetic photons from the sun are employed to dissociate water and thus to gain access to atoms of hydrogen. The hydrogen is next used to "reduce" (take oxygen away from) the molecule of carbon dioxide. This makes available unoxidized carbon, which can then be used for construction materials and for a variety of functional materials such as proteins. Unoxidized carbon can also be used to construct the various sugar-like substances (saccharides and polysaccharides) that provide storable sources of chemical energy.

When the photosynthetic organism dies, and when the other organisms that have benefitted from its products die, microbial decay will return to the atmosphere all the materials that have been taken out. Depending on the type of microbe undertaking the decomposition, the carbon will be returned to the atmosphere either as carbon dioxide or as methane (CH_4). Because the atmosphere is rich in oxygen, any methane released into it will spontaneously transform into carbon dioxide and water on a time scale of about ten years. So far as the energy balance is concerned, no chemical energy derived from the earth has been used up. Carbon dioxide returns as carbon dioxide, and water returns as water.

It may thus seem that carbon cycles through the surface biosphere in a complete and closed manner. If the atmosphere and the exposed rocks initially possess the volumes of raw materials required by life, the process should go on for as long as the sun shines and temperatures allow water to remain in a liquid state. But as we will see in Chapter 4, the path that carbon follows through the cycle of photosynthesis and oxidation is far from a closed loop. Several times as much carbon as is taken up by living materials is constantly extracted from the atmosphere and taken out of circulation, as long-lived or permanent carbonate rock. The surface biosphere must therefore have been kept alive by an ongoing and large supply of carbon in the form of either methane or CO_2 (or, as some observations would indicate, by a mix of the two). CO_2 will be the final addition to the atmosphere in either case.

In the surface biosphere, all the energy driving biochemical transformations ultimately comes from sunlight. Life in the deep hot biosphere does not have access to sunlight, so the source of energy could not work in the same way. But even there, carbon is the basic building block of life. What is the source of this carbon in the subsurface realm?

The notion, derived from surface biology, that CO_2 is the standard carbon supply for all life has been applied by some investigators to the deep life also. While the ocean water contains plenty of CO_2, it does not have any energy source to reduce this. The reduced carbon that trickles down from the surface layers would be quite inadequate. No energy can be derived from a process that both starts and ends with oxidized carbon. If unoxidized carbon were available at the outset, in the form of hydrocarbon molecules migrating upward, then these molecules would be the logical candidate for a carbon supply that would also yield an energy-producing sequence, ending up with CO_2.

The hot ocean vents are not themselves provinces of the deep hot biosphere; they are borderlands between two worlds, between surface and subsurface. Nevertheless, their food chains are driven by processes so different from that of the surface realm that they are a good place to begin our explorations of deep hot biosphere energy. The amounts of carbon that sink down from ocean surface life are quite inadequate to supply the exceptionally fertile ocean vent biology. The volcanic rocks of the sea floor contain only a very small fraction of carbon—about 200 parts per million (ppm). To extract carbon from this source would be difficult and very energy-consuming. There is, however, a much larger carbon source in all these communities: hydrocarbons. Methane (CH_4) is generally the most abundant, but the heavier members of the series, such as ethane (C_2H_6) and all the way up to oils constituted of twenty to thirty carbon atoms, are also found along the same fault lines, though in regions where less volcanic heat is in evidence. As the next two chapters will show, these hydrocarbon fluids show many features that suggest they have come up from much deeper regions.

The chemical energy supply, we might then suspect, is driven by the oxidation of these hydrocarbons. Starting out with hydrocarbons avoids the first and energetically most demanding step in the surface energy cycle. The chemical energy that is made available at the ocean vents is very similar to that made available by burning natural gas (which is largely methane) and turning it into water and carbon dioxide. There is one snag, however. When methane is burned in a furnace, there is an unlimited amount of oxygen from the atmosphere available all the time. In the ocean vents, a borderland between the surface and the deep biospheres, there may be some atmospheric oxygen available that was carried down in solution in the cold ocean water. If this were sufficient for converting all the methane supplied from the vents into carbon dioxide and water, then this borderland province would be dependent on surface biological processes, and it would not be an outpost of what I suggest is an independent realm of life stretching down into the rocks below. It seems doubtful that the prolific life at these concentrated locations on the ocean floor could receive enough water-borne atmospheric oxygen, but a firm answer is not yet known. However, this issue is not of central importance. We now know of many cases where we can probe so far down into the deep biosphere that atmospheric oxygen has absolutely no access, and we observe generally similar metabolic processes taking place there. Where does the necessary oxygen come from?

There is plenty of oxygen bound in the rocks, as noted earlier, but most of it is so strongly bound that more energy would be required to remove it than could be derived by using it subsequently to oxidize hydrocarbons. There are just two common substances in which oxygen atoms are bound loosely enough that more energy would be obtained from using oxygen so acquired than is spent in acquiring it. These two common substances are highly oxidized iron (Fe_2O_3 and associated compounds) and oxidized sulfur (such as SO_2 and H_2SO_4 in compounds that are called sulfates). If microbes at or beneath the ocean vents secure their oxygen needs from ferric iron oxides, what will remain is a less oxidized form of iron—magnetite or greigite. Microbial action leaves a clear fingerprint behind: The crystals of these products are much smaller than those of the same substances that have frozen out in the cooling of rocks from liquid to solid form.

The water of the oceans includes the second source of lightly bound oxygen, sulfate, in great quantities. Sulfate (SO_4) is the second most abundant ion of negative charge in seawater. The amount of oxygen that could be derived from marine sulfate ions may well exceed the con-vected atmospheric oxygen available at the ocean vents. If oxygen is, in fact, primarily available near the vents in the form of sulfate, then the microbes that make use of the hydrocarbons will be in an ideal situa-tion: The chemical transformations for extracting the chemical energy from upwelling hydrocarbons will not run by themselves, because an initial energy supply is required for the first step of freeing oxygen atoms from sulfate. The microbes will be amply compensated for this energy-demanding step, however, when the second step is taken.

The task of brokering such transactions is left to the world of microbes. Here, it is important to remember that a chemical fuel is use-less to life if it combusts spontaneously. Dinner would do you no good if the food burst into flames on your plate. For a substance to qualify as "food," it must become oxidized only with the help of a catalyst cre-ated and deployed by life. This is a fundamental requirement both for the organisms at the base of the food chains of the surface and deep biospheres and also for all organisms that stand later in line.

The removal of oxygen from sulfates at the ocean vents would pro-duce either pure (elemental) sulfur or sulfides, which are unoxidized sulfur compounds. The large quantities of metal sulfides that are found heaped up at the edges of ocean vents suggests that such biologically facilitated transformations are indeed taking place.

A further requirement for the construction of organisms—be they inhabitants of the surface biosphere or the subsurface biosphere—is a supply of various metals required in the protein molecules known as enzymes that catalyze chemical reactions. Also required for biological construction or chemical processing are some reactive molecules that contain elements such as sulfur, phosphorus, and chlorine. The required quantities of these are small enough that the upper crust of the earth can usually supply them. The deep biosphere and the land por-tions of the surface biosphere are thus adequately nourished. But the surface waters of the open oceans may be impoverished, particularly with respect to phosphorus and iron.

In summary, there are important differences and important similar-ities between the two biospheres. The surface biosphere runs on solar energy converted into chemical energy; the deep biosphere begins with chemical energy freely supplied from the depths of the earth. Both biospheres rely on unoxidized carbon as the building block of life, but surface life extracts it initially, with the help of sunlight, from carbon dioxide in the atmosphere, whereas deep life extracts it from the same substances used as the energy source: hydrocarbons. Oxygen is a requirement in both realms, since chemical energy is provided only in the process of oxidation. For surface creatures, oxygen is available largely in the form of pure, molecular oxygen. Inhabitants of the sub-surface must work harder to gain their supply, extracting oxygen atoms that are loosely bound in iron oxides and sulfates.

The Ecology of Deep-Ocean Vent Life

Because we are surface creatures, we are inclined to regard an ecosystem based on chemical energy rather than photosyn-thetic energy as a strange, if wonderful, adaptation of life. We marvel at the ecology of the deep ocean vents as a deft adjustment of surface life to an inhospitable realm. The evidence argues otherwise. Microbes and even animals are thriving at these vents; growth rates are thought to exceed those in even the most productive surface realms. If the theory of the deep hot biosphere is correct, we would infer that the microbial pioneers invaded from below. Many viewpoints would have to be changed as a consequence.

The communities of life at the deep ocean vents differ from other marine ecosystems not so much in their garish macrofauna but in their unseen microbes—the bacteria and archaea at the base of the food web. Two decades of studies have revealed that these microbes feed on mole-cules gushing from the vents: hydrogen (H_2), hydrogen sulfide (H_2S), and methane (CH_4), each of which can supply energy only if oxygen is avail-able.[1] No known animal can feed on any of these chemicals directly, but animals can feed on microbes that do. What is particularly remarkable about the deep-ocean vent communities is that many of the macrofauna seem to be dependent on symbiotic partnerships with the microbes.

Clams and mussels have entered into symbiotic partnerships with microbes bound in their gill tissues. The giant tube worm species, however, has taken partnership to a new dimension. Its interior guests are so skilled in producing food for themselves and their host that coevolution has atrophied the worm's digestive system and deprived it of a mouth. Utterly dependent now on the excess production of its symbionts, the tube worm has evolved a large and specialized organ deep inside its body for the microbes to inhabit. The worm supplies its microbes with the materials they need by employing feathery red gills to filter useful molecules out of seawater. Then it volunteers its own circulatory system to deliver what the gills have gathered.

The greatest challenge to organisms along hydrothermal vents is posed by the risk of being swept out of range of the vent and thereby losing the chemical supplies and the temperature range they require. The bivalves and tube worms solve the problem by anchoring themselves in place. Crabs and shrimps and snails that live among the fixed organisms can, of course, creep and clutch as needed. The microbes that constitute the primary step in the food chain have found ways to hold their place, too. The most heat-adapted varieties can live very close to (and even inside of) the vent. Wherever it is too hot for animal grazers such as snails to intrude, microbes cling to the rocks in communal mats of slime. Those that take to the water column above the venting fluids possess a whip-like flagellum by which to locomote, sensing temperature or chemical stimuli to guide their directional movements and thus staying within or next to the vent stream. The most audacious bacterial entrepreneurs are those that have made themselves welcome guests within the very tissues of the bivalves and tube worms. There they are protected from prowling grazers as well as errant currents.

The hydrogen, hydrogen sulfide, and methane fuels consumed by both free-living and symbiotic microbes in the vent communities are exploited by microbes that access oxygen atoms loosely bound in ferric iron oxide carried up from the depths in vent fluids, oxygen derived from sulfate that pervades seawater, and perhaps also free oxygen in the seawater.

All animals, however, depend on molecular oxygen for their metabolic needs. No animals are known to extract oxygen directly from oxidized materials in their surroundings. Many investigators have therefore assumed that the macrofauna at the vents depend on the molecular
oxygen carried down in seawater. Thus these species would still be
dependent—indirectly—on surface photosynthesis. They would still
be members of our food club.

The great abundance of molecular oxygen in the atmosphere is
mainly due to its production as a waste product of photosynthesizers—
by plants on land and algae and cyanobacteria near the surface of the
sea. Molecular oxygen diffuses into surface waters, especially at high
latitudes, because the solubility of oxygen in seawater is greatly
increased at low temperatures. Oxygen-rich waters from the Arctic and
Antarctic plunge to the deeps and then slowly snake along the ocean
floor, following valleys, toward the equator. A global-scale system of
atmospheric and oceanic circulation thus brings molecular oxygen to
some deep areas of the ocean floor.

Most of the ocean vents that have been discovered are situated at
volcanic ridges and high spots of ocean floor, and such areas do not
receive the cold, oxygen-rich flows of polar waters. Whether the oxygen that had diffused to these locations and is made available only by
slowly moving water would be sufficient to foster the extremely rapid
growth observed at the vents is doubtful. Although the macrofauna
cannot extract oxygen from other sources, microbial life can. If the supply of oxygen is the limiting factor for the vent community, then we
have to suspect that symbiotic exchange may have advanced to such a
state that the symbiont microbes within the animals are stripping oxygen atoms from seawater sulfate not only for themselves but also for
their animal hosts. No doubt further researches will determine whence
the macrofauna derive their oxygen. But, as we shall discuss in the
next section, many microbial communities have been identified that
certainly have no access to atmospheric oxygen.

Life thrives at the ocean vents because these are sites at the borders
between two worlds. An abundance of chemical energy can be
extracted from the chemicals that meet there and that had no opportu-

nity to reach equilibrium with one another. Upwelling fluids from the world below are rich in "reduced" molecules, such as hydrogen and methane. Hydrogen sulfide is also present, but we do not yet know whether this is a primary fluid from the depths of the earth or a product of microbes as they utilize a combination of hydrogen and sulfate for energy needs.

Of the three major sources that provide energy when reacted with oxygen (hydrogen, hydrogen sulfide, and methane), hydrogen sulfide has attracted the most research interest, because it seems to be the fuel on which the microbial symbionts of the giant tube worms and clams depend. But the carbon atoms that form the core of all organic molecules must be obtained elsewhere. The presence of methane in the output of ocean vents thus assumes particular importance; it can be the source of the required carbon as well as the source of chemical energy.

Hydrocarbons bear a structural resemblance to foods we eat that are derived from photosynthesizers. For example, the only material difference between a molecule of hexane (a six-carbon form of petroleum) and a molecule of glucose (a six-carbon sugar, common in foods at the surface) is that hydrogen atoms surround the chain of carbon in hexane, whereas water molecules surround the chain of carbon in the sugar. The hexane C_6H_{14} is a *hydrocarbon,* whereas the sugar $C_6H_{12}O_6$ is a *carbohydrate.* The terminological difference is subtle but important. For us animals the carbohydrate is food, the hydrocarbon poison. Nevertheless, the biological idiosyncrasies of our own tribe of complex life should not be allowed to constrain our judgment as to the possibilities—indeed preferences—among the multi-talented microbes. They might well have a metabolism that requires an input of petroleum.

Microbes that utilize methane as a source of energy in the presence of oxygen, and also as a source of carbon, are known to be present in the hydrothermal vent communities. Such *methanotrophs* ("methane eaters") have been identified as symbionts within the macrofauna—thus far, only in mussels—but they are presumably free-living as well.[2] They can consume heavier hydrocarbons, too.

Are the methanotrophs of the deep-ocean vents ambassadors from this other, deeper, and perhaps independent world? We know that

clams and worms do not venture any deeper than the thin skin of surface rock and sediments. But what about the bacteria and archaea? If
microbial slimes on the rocks near and within the vents thrive on
methane and sulfide gases that rise up from below, might they not also
find suitable habitat within cracks and pore spaces deep below the
crustal surface?

Other Borderland Ecologies

W ithin the past three decades, many and
various borderland ecosystems have
been discovered and their secrets
probed. First to capture scientific attention was a type that had long
been enjoyed by crowds of tourists: the microbial communities that
colorfully coat the rocks within hot pools of Yellowstone National
Park. Serious study of the metabolisms of Yellowstone's thermophilic
(heat-loving) microbes began in the mid-1960s.[3] It was here that scientists first came to appreciate the extraordinary talents of the earth's
seemingly simplest forms of life. For example, one bacterium discovered in Yellowstone's hot pools, *Thermus aquaticus,* provided the
enzyme that launched the molecular biology industry by making DNA
replication fast and easy. Today, Yellowstone's hot springs offer rich
prospecting for scientists seeking to add new names to the list of
microbes classified in the taxonomic domain of Archaea.

In 1977 the exciting exotica we have already discussed were discovered beneath the sea—the elaborate assemblages of microbes and animals at the edges of hot springs on the ocean floor. In 1984 came the discovery of more assemblages of symbiotic microbes, tube worms, and
bivalves—not, this time, in the abyssal depths but on the much shallower continental shelves.[4] Similar in form, but taxonomically different
at the species or even genus level, tube worms and bivalves on the continental shelves were making their living in "cold seep" regions, where
crude oil and hydrocarbon volatiles seep up through the sediments. No
hot springs or other hydrothermal action is associated with these seeps.
Unlike the hydrothermal vents, which are point sources restricted in

size, cold petroleum seeps offer marine life chemical energy over vast expanses of the continental shelves that are too deep to support photosynthesis. (In even the clearest ocean waters, photosynthesis is impossible any deeper than about 200 meters beneath the ocean surface, and continental shelves often sink to a depth of a kilometer or more.) Growth rates in the regions of hydrocarbon seeps are not, however, as high as they are at the actively venting rift zones of the deep ocean.

On land, too, an ecosystem border realm has captured scientific and public attention. In 1986 a cave in Romania—until then, isolated from the atmosphere—was discovered and found to contain a thriving ecosystem based on the chemical energy of reduced gases emanating from below. Ten years later, when its biological inventory was published, this cave habitat was touted by the media as the first instance of a terrestrial ecology that was not based on photosynthesis and yet was able to support not just microbes but land animals as well.[5] Feeding on the bacterial base of the food web are more than forty species of cave-dwelling invertebrate animals, including spiders, millipedes, centipedes, pillbugs, springtails, scorpions, and leeches. Thirty-three are new to science. As with the deep-ocean vent habitat, hydrogen sulfide was identified as the reduced gas supporting the base of the food chain in this cave, though I suspect that methane also plays a role. Indeed, methane consumers may well be generating hydrogen sulfide as a waste product when sulfate is used to oxidize methane, in which case the sulfide consumers would be a notch up from the base of the food chain. Hydrogen sulfide, converted by water into sulfuric acid, probably carved out the limestone cave.

In 1997 another cave ecosystem based entirely on chemical energy was explored in southern Mexico. That cave, too, appears to have been carved out of limestone by a flow of sulfuric acid. The acid fumes in this cave are so intense that scientists were able to venture a mile into its tunnels only with the assistance of breathing masks. Microbial life is so prolific throughout that the walls are shrouded in slime.[6] Feasting on the microbes is a community of invertebrates, but this ecosystem also supports vertebrates: tiny fishes in the waist-deep water that occupies the tunnel system.

Also very recently, Russian scientists have been preparing to explore a vast lake—as large as Lake Ontario—that was discovered in central Antarctica beneath four kilometers of ice.[7] Lake Vostok owes its existence to the entrapment of heat upwelling everywhere within the earth. The thick glacial ice, strangely enough, acts as a thermal insulator, segregating the heat from the intense cold of Antarctic air. Remote sensors indicate a water depth of perhaps 600 meters in some places, underlain by sediments 100 meters thick. Drilling was halted 250 meters above the water line, pending implementation of procedures that could ensure sterile contact. If life is present down there, it will unquestionably be based on chemical energy welling up from below. To test that possibility, it is imperative to prevent contamination of the pristine lake by surface microbes. NASA has expressed interest in fostering technologies for sterile exploration of Lake Vostok, which would probably happen no sooner than 2001. One reason for NASA's interest is that a subglacial lake offers an extraordinary analog for the subsurface environment of Europa, a moon of Jupiter that is covered with a thick layer of ice and may have liquid water underneath that.[8]

An important discovery of very large amounts of methane was made in the last two decades. Methane hydrates, crystals of water ice that entrap methane molecules within their lattices, exist in great quantities on many areas of the ocean floor. The presence of methane raises the freezing point of water by an amount depending on the ambient pressure, and therefore this ice can form in regions where water is supplied in liquid form and then freezes where methane is added.

For methane hydrates to form, temperatures must be no greater than about 7°C and pressures no less than about 50 atmospheres. This means that much of the sea floor that is outside of volcanic zones and covered by water to a depth of 500 meters or more could support methane hydrates.[9] Within the past two decades we have learned, both by remote sensing and by direct sampling, that methane hydrates do indeed exist in great quantities in many areas of the ocean floor. They produce a clear and unique signature on sonar and are remotely sensed as a distinct layer in ocean muds, sometimes lying directly on the bedrock of the ocean floor. A large area of the continental shelf has been surveyed in this fashion. Results indicate that methane hydrates may, in fact, be present in all areas where the pressure and temperature allow them to form.[10] It has been estimated that methane hydrates (those within the Arctic permafrost layer as well as those under the sea) contain more unoxidized carbon than all other deposits of unoxidized carbon known in the crust, such as crude oil, natural gas, and coal.[11]

Often there is more carbon in the methane atoms trapped in a deposit of hydrate than in all of the sediments associated with that deposit. In such instances the conventional explanation of its source (biological materials buried with the sediments) cannot account for the production of so much methane. The methane embedded in the ice lattices must have risen from below, through innumerable cracks in the bedrock. Once a thin, capping layer of the solid forms, the genesis of more such hydrate underneath becomes an inevitability, provided methane continues to upwell.

This conclusion—that the source of methane lies beneath, not within, the crustal sediments—is strengthened by evidence of pockets of free methane gas beneath some regions of hydrate ice[12] and also beneath permafrost layers of Arctic tundra.[13] In these regions, downward migration of methane gas from overlying sediments does not seem conceivable. Gases, after all, do not migrate downward in a liquid of greater density. If there is any flow, it is in the reverse direction.

Lake Vostok, which we have just discussed, will be an ideal place to check on the quantity of hydrocarbons that have come up from below since the ice cover formed. The quantities of methane hydrates contained there may be very large, they may even represent the major component of the lake.

In the domain of high methane hydrates there is also macro-life, just as at the ocean vents. Little worms are found there that plough through the methane hydrates and the overlying water.[14] Their existence indicates that such methane hydrates have been there long enough to allow life to adapt to the strange circumstances. Most probably symbiotic microbes inside the worms use energy derived from the oxidation of methane. The carbohydrates and other biological molecules the microbes produce are then shared with their animal hosts.

Hydrates made up with CO_2 rather than methane can exist also, though over a smaller stability range of temperature and pressure than methane hydrates. Nevertheless, there are substantial areas of ocean floor that could support CO_2 hydrates, but few–if any–such samples have been found. The conclusion must be that the "gentle" but widespread addition of carbon to the atmosphere is a global phenomenon of diffusion from the ground of methane and other hydrocarbons, no doubt at different rates at different locations and at different times. The dominance of CO_2 over methane from volcanoes is the exception and not the rule. This conclusion then agrees with the finding that methane is far more abundant than CO_2 in wellbores (to the good fortune of the petroleum industry), and also with the evidence from meteorites that hydrocarbons and not CO_2-producing compounds will have been the principal input of carbon in the forming earth. (Chapters 3 and 4 will explore these points in detail).

Deep Is Desirable

In light of the discoveries of thriving chemical-based ecosystems associated with methane hydrates, hot ocean vents, and cold petroleum seeps on the ocean floor, along with those associated with hot springs and gas-rich caves on land, we can conclude that methane, hydrogen sulfide, and other energy-rich gases (those that could provide large amounts of energy if combined with supplies of oxygen) are attractive to life forms that span a wide range of temperature. Very close to the hot ocean vents, however, and wherever hot springs on land are more than merely warm—above, say, 45°C—these habitats do not support animals. But heat-loving (thermophilic) microbes are abundant in these places.

As temperatures rise even more, thermophiles drop out, but hyper-thermophiles—microbes that grow best at 80°C or higher[15]—go about their business unperturbed. The waxy cell membranes characteristic of hyperthermophiles facilitate material exchange at temperatures at which fatty membranes like our own would simply melt.[16] Hyperther-mophiles can grow and reproduce only at such high temperatures. At lower temperatures their membranes stiffen to the point where materials can no longer pass through as needed. Molecules called heat-shock proteins enshroud the DNA and regular proteins of hyperthermophiles, guarding the intricately folded structures against the unraveling that such high heat would otherwise bring about.

What are the highest temperatures that hyperthermophiles can tolerate? We are still uncertain. But we do know that temperature alone is no more determinative of an environment's livability than it is determinative of a fluid's boiling point. One more factor must be considered: pressure.

Although the boiling point of water is 100°C at sea level, it rises to a full 300°C at a depth of just 876 meters. At that depth, the water column exerts a pressure of 87 atmospheres, which means 87 times more than the pressure exerted by the atmosphere at the surface of the sea. This pressure is sufficient to prevent water molecules at even 299°C from expanding into a vapor phase. Deeper still, at a depth of 2.25 kilometers, the "critical point" is reached. Here the pressure is so great that no matter what the temperature, there is no longer any distinction between vapor and liquid. Rather, it is more appropriate to refer to water beyond the critical point as existing as a fluid—specifically, a "super-critical" fluid.

Now consider that the first community of hydrothermal vent organisms ever witnessed in the abyssal realm of the sea was found at a depth of 2.6 kilometers. Here water is a supercritical fluid. Water at temperatures of about 300°C has been detected issuing from the vents, but it is cooled quickly as a result of mixing with surrounding water. Boiling is not an issue for organisms at that depth, because water cannot boil there. Melting of membranes and unraveling of proteins, rather, may become the limiting factors for life at high temperatures.[17]

Because of the effect of pressure, if one must cope with temperatures approaching or exceeding 100°C, then deep is certainly desirable. How widespread are zones of such high temperature? Hot springs— whether on the sea floor or on land—are far from the norm. They occur where heat generated deep within the planet finds a rapid escape route to the surface, by way of fluids buoyed up from below. These are active volcanic zones. Far more common are non-volcanic regions, such as those over which you and I are probably sitting right now.

The earth generates its own heat from compression, gravitational sorting, and radioactive decay deep within its core and mantle. In a non-volcanic region the temperature of the rock, beginning at the surface, increases steadily with depth and at a rate fairly uniform over the entire globe. This phenomenon is referred to as earth's thermal gradient. The temperature of the crust near its contact with the atmosphere is approximately 20°C over most of the area. The temperature increases at a rate of between 15°C and 30°C per kilometer of depth in non-volcanic regions.

Hyperthermophiles are known that can grow at temperatures of 110°C. This means that, on average and provided that the necessary chemical resources are present, life as we know it could survive down to a depth of six kilometers in regions of crust that exhibit the low temperature gradient (15°C per kilometer or less) and three kilometers where the temperature gradient is high (30°C per kilometer). It is not yet clear whether hyperthermophiles exist that can tolerate higher temperatures still. Some microbiologists consider that the temperature limit for microbial life may be as high as 150°C.[18] In that case, life might extend to deeper levels, in some cool areas possibly to a depth of ten kilometers.

It is crucial to remember that because of the steady rise in pressure with depth, nowhere within the earth's crust (with the exception of volcanic zones) does the combination of temperature and pressure ever permit water to boil. What about methane, the lightest and hence quickest to boil of all hydrocarbons? Moving downward along any thermal gradient, methane becomes denser at the greater pressures of increasing depths, even as it remains a vapor. What does this increase in density mean for subterranean life forms that feed on methane?

For one thing, the greater density means that methane is actually easier for life to access at depth. At a depth of six kilometers, for example, methane would be 400 times as dense as it would be on the surface at atmospheric pressure. Also, higher temperatures that coincide with greater depths escalate the rate at which methane molecules collide with the cell membranes of microbes. Both factors enhance the rate at which methane would be expected to diffuse across cell membranes.

Deep is thus desirable not only to ease some of the biological problems created by high temperatures but also to assist methane consumers in accessing their food.

Up here in the surface biosphere, where methane exists only as a diffuse gas, methane consumers are a curious group. But methanotrophs may be far from tangential members of the food web in the deep biosphere. Indeed, they may be the foundation of that system.

Beneath the Borderlands

To study the deep hot biosphere and sample its inhabitants, we must probe far beneath the borderland regions of hot springs, hydrothermal vents, oil seeps, methane hydrates, and gas-rich caves. We must peer into the bottom of deep wells drilled into the earth's crust.

When I began developing the deep hot biosphere idea in the early 1980s, and when my "Deep, Hot Biosphere" paper was published in 1992,[19] a persistent criticism was that microbes brought up in samples from the depths of oil and gas wells were not native inhabitants but opportunists introduced from the surface in biologically contaminated drilling fluids.[20] This contamination argument was at first difficult to refute. But in 1995 a key paper published in one of the top scientific journals demonstrated that microbes discovered at a depth of 1.6 kilometers in France were truly "members of a deep indigenous thermophilic community."[21] The following year another report of indigenous microbes, this time from an oil well in Alaska, established active biology at a depth of 4.2 kilometers and a temperature of 110°C.[22] In 1997 the discovery of microbial *fossils* embedded in granitic rock at a depth of 200 meters confirmed the indigenous interpretation; fossils cannot be introduced by drilling fluids into solid granite.[23]

Thus far, the deepest indication of active biology was detected in 1991, at a depth of 5.2 kilometers in Sweden, as we will see in Chapter 6.[24] Significantly, the well in which these microbes were detected had been drilled into solid granitic bedrock, not the sedimentary strata that generally attract petroleum prospectors. A sample that had been taken and sealed at depth and then drawn up was cultured in the laboratory. It yielded previously unknown strains of anaerobic microbes that reproduced only in the temperature range from which they had been sampled, 60°C to 70°C.

The term I coined, *deep hot biosphere,* is sometimes mentioned in scientific papers or media coverage interpreting such findings of microbial life discovered at depth.[25] Many of these reports, however, misunderstand my argument and, I believe, misinterpret the facts in ways that are far from trivial. These errors are of two types.

First, microbes drawn from deep oil wells are rightly interpreted as feeding on hydrocarbons. But there is an implicit assumption that the hydrocarbons are the reworked remains of life that once belonged to the photosynthetic food club—algae and the like.[26] This is the standard Western view of a putative biogenic origin of petroleum, which I will challenge in the next chapter. As long as petroleum is regarded as biogenic, then no matter how far down life may be found in oil wells, it will always be regarded as a novelty—a thrilling extension of the surface biosphere downward as it mines its own earlier remains.

A second error in reports heralding the discovery of a deep biosphere, or even a deep hot biosphere, is the characterization of indigenous microbes as "rock-eating." This second error requires a bit more explanation than the first. To begin with, "rock-eating" is the usual interpretation of microbial metabolism when microbes are discovered in wells drilled in igneous rock. Because igneous rocks formed from a melt, the only hydrocarbons they could possibly contain must have migrated from somewhere else after the magma cooled into rock. The standard way of thinking would have those hydrocarbons seep into the cracks and pores of igneous rock from a sedimentary "source" rock (such as black shale) nearby. When there is no nearby source rock, this explanation is of no use.

Reports of microbial life within igneous rocks are considerably less widespread than reports of microbes detected in sedimentary rocks. The reason for their scarcity is simple: If we believe that oil and gas are the reworked remains of surface life long buried in sediments that consolidated into rock, then why would anyone drill in igneous territory?

The number of boreholes drilled in sedimentary rock is so much larger than the number drilled in igneous rock that this disparity alone can readily account for the difference in the number of reports of microbial life from the two domains.

Deep drilling into non-sedimentary rock has nevertheless been undertaken, either for explorations of a general kind or for an altogether different purpose: to assess the radioactive contamination of ground water. During the Cold War, radioactive wastes generated by plutonium production were not always disposed of carefully. This was the case at the Hanford nuclear processing facility in central Washington state, which was built on Columbia River basalt. A test well drilled 400 meters into the igneous rock to sample radioactive contamination of aquifers had the side effect of revealing bacteria.[27] What were they living on? Because everyone believed that such an extensive basalt could not possibly contain hydrocarbons, the plentiful supply of methane detected there[28] was interpreted as a metabolic by-product of a later stage in the food chain (with what source of carbon?)—rather than, as I would have it, the fuel source for the primary producers.

In igneous rocks, methane is by far the most common fluid, second only to water. Methane is the most likely fuel source, carbon source, and hydrogen source at the base of the food chain. To my way of thinking, carbon dioxide is largely a *product* of microbial oxidation of hydrocarbons, not the *source* of carbon for the base of the food chain. This view—that hydrocarbons provide the carbon source as well as the fuel for biosynthesis at depth—has been greatly strengthened by a paper published in 1994.[29] Petra Rueter and colleagues cultured a moderately thermophilic microbe in conditions that confirmed that this metabolic strategy was in use, with sulfate providing the oxidant.

For many reasons, therefore, I do not agree with the ecological interpretations of the researchers working on the Columbia basalt aquifer. Nevertheless, I can well understand how misinterpretations could have been made. It is difficult to sample, culture, and identify the presence of indigenous life at depth. It is even more difficult to determine the foundation of the food web and the fuel and material sources on which the primary metabolism is based. Until the primary metabolism is identified, however, one cannot be sure whether a particular chemical constituent is original resource or biological product.

It should now be clear that the best way to learn about the deep hot biosphere—and even to test whether this hypothesized realm of life does indeed exist—is to drill into rocks and examine what is down there. Few if any holes have yet been drilled with the express aim of searching for deep life. Wells are drilled to search for commercial quantities of hydrocarbons, to test for contamination of ground waters, or to provide data for understanding geological processes. Any microbial life encountered during such ventures is almost always dismissed as biological contamination from the surface, introduced in the drilling fluids.

The borderland habitats are exciting, but they cannot demonstrate with certainty whether and what biological processes may be active at depth. Thus far, we have had only glimpses of what may prove to be a vast expression of earth life awaiting our exploration. There has, fortunately, been a recent surge in demand to study microbes hauled up from depth. Interest in a deep hot biosphere (though not necessarily my stringent interpretation of an independent, hydrocarbon-based deep biosphere) has blossomed. Part of this interest has been stimulated by the numbers of deep wells that have tested positive for biological inhabitants. Life is not supposed to be down there, so our curiosity is piqued. Another substantial part of the interest is attributable to the success of the University of Illinois evolutionist Carl Woese in convincing biologists that a whole new domain of life awaits exploration— the Archaea.[30]

Until recently, all living organisms were classified either as prokaryotes (which included all that was then called bacteria) or eukaryotes (which included plants, animals, fungi, slime molds, and single-cell protoctists). Several important cellular features distinguished these two groups, the principal difference being that, in contrast to eukaryotes, prokaryotes lack a nucleus to hold their genetic material.

In the 1960s and 1970s, Woese discovered that one set of prokaryotes (which he called archaebacteria) was vastly different from all the rest in its ribosomal RNA sequences and in some important metabolic and morphologic features. When the full genetic sequence of one of these aberrant microbes was published in 1996,[31] it was clear to most experts that the prokaryote classification would have to be rethought: The archaebacteria contained many unique genes, and archaebacterial genes seemed to have more in common with those of eukaryotes than with those of the rest of the prokaryotic tribe. As a result of this work, the prokaryotes are now typically considered two "domains," one still known as Bacteria (or Eubacteria), the other called Archaea. (Eukarya has remained as before, but it is now regarded as the third domain in this taxonomic system.)

Woese's conceptual revolution highlights the importance of undertaking further research in the domain Archaea, about which so little is known. Moreover, the three-domain classification of life indicates that hyperthermophily is the most ancient of traits.

The reclassification of microbial life proposed by Woese has a strong resonance with the concept of a deep hot biosphere. The separate branch of life that he has called the Archaea must have had an early origin in the evolution of life, judging by these organisms' simple genetic systems, and because so many strains are hyperthermophiles, they must have originated at a high temperature. It seems very improbable that one form of the thermophilic Archaea developed on one hot ocean vent and spread from there to many other locations, evolving into the great variety of strains we now observe. It seems more likely that they represent a global evolution of an early form of life that depended on the supplies of chemical energy that the earth delivered. Archaea would thus be the product of a long evolution in a large, connected, and long-lived habitat. They may be the earliest inhabitants—and even today the principal inhabitants—of the deep hot biosphere that embraces the earth.

To probe the origin of earthly life, we must look to the organisms that thrive in extreme heat. Only the bacterial and archaeal domains include hyperthermophiles, and only Archaea is dominated by this trait. Where do we look for hyperthermophiles? We can, of course, probe the borderland systems of hot springs and deep-ocean vents, but by far the greatest opportunity surely resides within the further reaches of the deep hot biosphere.

The deep hot biosphere theory will ultimately stand or fall on evidence gained from the *deep* earth—not from the borderlands, although the borderlands do indicate the richness of possibilities, as in the case of deep-ocean vents, cold petroleum seeps, methane hydrates, and gas-rich caves. Among the most interesting aspects of hydrothermal vent communities is that the methane involved in life support in that realm has no biogenic explanation, and yet methane is known to be abundant in hydrothermal vent fluids.

A biogenic explanation is implausible because there is little surface-derived sediment on, below, or even near the active rifting zones of the deep ocean floor. All rocks as far down as one may wish to probe are igneous, which means they once flowed upward through the crust as exceedingly hot liquid magma. Yet in the popular opinion we all learned in school or picked up from our culture, natural gas and petroleum are regarded solely as the remains of organisms reworked by geology into "fossil fuels." Elevated (but not volcanic) temperatures and elevated pressures—both induced by burial of organic materials that were once alive—will, given sufficient time, perform this feat of alchemy within the sediments that contain organic residues (or so we are told). Methane is therefore something to be found in sedimentary rocks, not igneous regions. How, then, can methane be associated with the volcanic rifting zones of the ocean floor?

If the contention that the earth's stores of petroleum liquids and most gases are in no way derived from biology is correct, we should be able to find these hydrocarbons in igneous as well as sedimentary regions. Crude oil is not a stew of cooked algae, and though attempts have been made, it has never been synthesized in a laboratory from biological materials. Some methane is indeed biogenic. Microbes called methanogens ("methane generators," as distinct from methane-eating methanotrophs) live in oxygen-poor habitats, such as in the muds of rice paddies and the digestive tracts of cattle, where they produce methane as a by-product of their feeding strategies. We are also familiar with methanol gasoline supplements distilled from Iowa corn. But in my view, the greatest portion of methane by far is not biogenic. Rather, hydrocarbons must be understood as primordial constituents of solar system debris out of which the earth formed some 4.5 billion years ago. This crucial point will be explored in the next chapter.

Chapter 3 # The Deep-Earth Gas Theory

N ew ideas in science are not right just because they are new. Nor are old ideas wrong just because they are old. A critical attitude is clearly required of every seeker of truth.[1] But one must be *equally* critical of both the old ideas as of the new. Whenever the established ideas are accepted uncritically and conflicting new evidence is brushed aside or not even reported because it does not fit, that particular science is in deep trouble. This has happened quite often in several fields. In geology, for example, a person who thought that continents or parts of continents might have moved in the past was ridiculed before 1960, despite the existence of good evidence from magnetic rock measurements. After 1965 anyone who did not believe in such movement was again a subject of ridicule. In petroleum geology, the massive and persuasive evidence for a deep origin of the fluids is still treated with disdain and cannot be published in certain journals.

Carbon and hydrogen can form a great variety of molecules that have different ratios of carbon to hydrogen and different molecular geometries, and all are called hydrocarbons. At the temperatures and pressures on or near the earth's surface, some hydrocarbons are solid (coal), some are liquid (crude oil), and some are in the vapor state (natural gas, which is predominantly methane). Liquid and gaseous hydrocarbons are commonly called petroleum, which exhibits great variation in the proportions of the various hydrocarbon molecules. Petroleum also has unifying features that suggest a similar mode of generation. How, then, is petroleum formed?

At the present time, most petroleum geologists outside the former Soviet Union would say that the question has been completely answered—that deposits of biological debris, reworked by geological processes, account for all natural petroleum. Elevated temperatures (but not elevated to volcanic levels) and elevated pressures prevailing at depth will, given sufficient time, perform this feat of alchemy, transforming the remains of surface life buried within sediments—or so we are told. Petroleum is therefore regarded as "fossil fuel." Yet the assemblage of widely accepted facts on petroleum chemistries and their geographical and geological occurrences, considered as a whole, does not support a preference for this standard solution.

The alternative explanation, which I favor, is referred to as the *abiogenic,* or deep-earth gas, theory. In this view, natural gas and crude oil are derived not from biological debris but from the initial materials that formed the earth. The goal of this chapter and the next is to present the arguments and evidence for this view. The abiogenic theory, in turn, will then serve as the foundation for our discussion of the title subject of this book: the deep hot biosphere theory.

The Origin of Petroleum: Two Conflicting Theories

E ven though the biogenic origin theory leads to many inconsistencies (which will be addressed in Chapters 4 and 5), it is nevertheless now virtually impossible in the Western world to conduct any research in petroleum geology that implies a questioning of this accepted position. A young person—however brilliant—with no scientific standing who attempted to do so would have no hope of passing peer review either for obtaining funds or for publishing contrarious results. Fortunately for me, by the time I began nosing around in the field of petroleum geology, I had established a favorable standing for myself in the fields of physics, including geophysics, and in astronomy. I had by then been elected to memberships in several prestigious learned societies, and this standing made it possible for me to air my heretical views on the origin and ubiquity of oil and natural gas.

Beginning in 1977, I wrote a number of papers on the subject of "deep-earth gas," in which I explained my reasons for thinking that natural gas and other hydrocarbons had originated at great depth—perhaps 100 to 300 kilometers beneath the earth's surface.[2] This depth is nearly 100 to 300 kilometers deeper than proponents of the biogenic view would place the origin of petroleum, as a consequence of their central presumption that petroleum forms from the remains of surface life, buried with the sediments. I presented the deep-earth gas theory during the time of the so-called energy crisis, which, to my mind, had arisen not because there was a physical shortage of oil and gas but because a cartel of major oil producers had gained much strength when several senior petroleum geologists forecast that within fifteen years all the reservoirs of crude oil in the world would be exhausted. It was then in the interest of the oil producers to cut back on production and exact the most revenue possible from the remaining reserves. Now, twenty-five years later, the world is awash in oil and has more than it requires—even by conservative estimates and even projecting significantly increased rates of consumption.

My proposal (and that of many Russian colleagues) that petroleum is abiogenic and ubiquitous deep in the earth, though far from the mainstream of opinion, did receive attention—particularly from petroleum entrepreneurs[3]—because of its practical importance well beyond the boundaries of pure science. In 1982 I supplemented the deep-earth gas theory in my own mind with the concept that a "deep hot biosphere" was thriving on these deep resources. A full decade passed before I was able to publish this hypothesis.[4] In taking this next step, however, I finally managed to put together all the pieces of evidence—including some that had initially been supportive of the biogenic theory of origin—in a way that I felt provided a satisfactory resolution of all the paradoxical information.

The origin of petroleum has been the subject of many intense and heated debates since the 1860s, when crude oil was first discovered to be present in large quantities in the pore spaces of many rocks. Was it something present when the earth was first formed? Or is it a fluid concentrated from huge amounts of vegetation and animal remains that may have been buried in the sediments over hundreds of millions of years? Arguments have been advanced for each of these two viewpoints, and although they seem to conflict, each line of argument has its strong points.

The biogenic theory holds that biological debris buried in sediments decays into oil and natural gas in the long course of time and that this petroleum then becomes concentrated in the pore spaces of sedimentary rocks in the uppermost layers of the crust. The search for oil was conducted with this theory of biological origin in mind, so the presence of biological material in the sediments was regarded as a key indicator of strata worth prospecting. Where petroleum reservoirs were found in rocks possessing no materials that could have given rise to the oils, it was simply accepted that crude oil and natural gases often migrate vast distances and that source rocks may therefore sometimes be indeterminable.

The biogenic theory of the origin of petroleum was widely adopted around the 1870s, when the earth was thought to have formed as a very hot body, perhaps a body of molten rock. If this had been correct, then no hydrocarbons supplied with the hot rocky material could have survived; they would all have been oxidized to CO_2 and H_2O. So long as this mode of origin of the earth was the dominant view, an abiogenic origin of petroleum, formed from materials accumulated in the formation of the earth, was not a tenable viewpoint. At that time, the formation of petroleum from vegetation, after the surface had cooled sufficiently, seemed to be the only possible explanation. The subsequent discovery of molecules of clearly biological origin in all natural oils greatly strengthened the biogenic theory.

The present theory of the formation of the earth is that it formed by the assembly of cold solid pieces condensed from a nebula surrounding the sun. Much of the material so acquired would have escaped excessive heating, and an abiogenic solution now seemed possible; but the biogenic theory was by then so firmly entrenched that opposing evidence was brushed aside. Even when, in the 1940s, the presence of many hydrocarbons on other planetary bodies of the solar system was discovered (bodies that could not have acquired them from vegetation), it continued to be maintained that just our earth acquired hydrocarbons from a source that could be supplied only here: vegetation.

Now, whenever crude oil or natural gas is encountered in igneous rocks (rocks that froze from a melt), the hydrocarbons are deemed to have migrated from a sedimentary "source" rock. In this view, igneous rocks underlying the deepest sedimentary rocks offer no prospect whatsoever for containing hydrocarbons, and so very few holes have been drilled into these "basement" rocks. Nearly all wells were drilled in sedimentary rocks, so nearly all oil was produced from sedimentary rocks. Before long, this fact was then taken to show that sediments were essential for producing oil. Sedimentary strata were indeed essential for the production of much of the oil we now have, not because there is necessarily more oil in the sediments but because that is where oil companies chose to drill. Belief in the biogenic origin of petroleum thus led to a self-fulfilling prophecy.

The theory of the biological origin of hydrocarbons was so favored in the United States and in much of Europe that it effectively shut out work on the opposing viewpoint. This was not the case in the countries of the former Soviet Union. Much work has continued there, on both sides of the debate, since the middle of the nineteenth century. In attempting to resolve this issue, the Soviet Union seems to have been more lenient toward scientific dissent than were the Western countries, probably because Mendeleyev, the revered Russian chemist, had supported the abiogenic view. The arguments he presented are even stronger today, given the greatly expanded information we now have.

The abiogenic theory holds that hydrocarbons were a component of the material that formed the earth, through accretion of solids, some 4.5 billion years ago. With increasing internal heat, liquids and gases were liberated, and because they were less dense than the rocks, buoyancy forces drove them upward. In favorable conditions, the upward journey from the regions of origin would be dammed temporarily in porous rocks at depths that our drills can reach, and from which we then derive commercial petroleum.

In volcanic regions we have a different situation. There channels of liquid can extend to great depths without interruptions, as pressure differentials between the solid rock and the nearly equally dense magma will be small. If methane from deeper levels enters such a channel, it will ascend as a mass of bubbles, and each bubble will have contact with fresh magma surface many times over in the ascent. Whatever loosely bound oxygen may be available there will be captured by the bubbles and at the high temperature will oxidize the methane to CO_2 and water. So it is not surprising that the emission from volcanoes at quiet times produces mostly CO_2 and water, and only a small percentage of methane (reported in most volcanoes as 2–5 percent, but much higher in some; in the Azores the figure quoted is 17 percent). But in major eruptions of these same volcanoes often a large amount of flammable gas is involved, and flames have been seen on many such occasions. The most clearly identifiable case was in the course of eruptions under the sea surface of one of the Krakatau volcanoes in the Sunda Straits, eruptions that did not break through the surface of the water but resulted in flames dancing on the surface over large areas. In this case there can be no confusion between flames and volcanic spray of red-hot ash, as has been suggested for many events where the presence of flames had been reported. Seemingly reliable reports of flames have also come from Central American volcanoes, from Santorini in the Mediterranean north of Crete, and from the great African Rift. (The chance of seeing the flames in an eruption is dependent on wind driving the dense smoke aside from the more vertical flame.)

In a violent eruption there will not be the small bubbles that come up at quiet times; instead there will be large plumes of gas, racing upwards through the molten rock. The contact area between gas and rock will be much smaller, and the time of such contact much shorter, thus reducing the amount of oxidation that can take place. All in all, a variety of evidence indicates that hydrocarbons or hydrogen are major components of the volcanic gases.

The CO_2 that is commonly seen in volcanoes at quiet times gives no proof that CO_2 is the primary carbon gas supplied to the surface of the earth. Where the emission of gases into the atmosphere can be measured directly, methane is almost always the dominant carbon gas, except when the measuring zone approaches an area of active volcanism, and there CO_2 often dominates. (I will return to this point in the discussion of mud volcanoes in Chapter 8.)

Plumes of hydrocarbons that find their way to the earth's surface without encountering magma may or may not be oxidized en route. They will in any case be oxidized soon after exposure to the oxygen-rich atmosphere. What this means is that the ultimate fate of primordial hydrocarbons is to be oxidized into carbon dioxide and water.

The abiogenic theory of petroleum formation depends on the truth of five underlying assumptions. First, hydrocarbons, or compounds that could have been converted into hydrocarbons at the intense pressures of the earth's depths, must have been a common constituent of the primordial materials out of which the earth was formed. Second, in the four and a half billion years since the earth accreted, the primordial hydrocarbons must not subsequently have become dissociated and fully oxidized to carbon dioxide and water by exposure to the significant amounts of oxygen bound in the rocks of the earth's crust. Third, hydrocarbons must be chemically stable at the combinations of high temperature and pressure that prevail deep within the earth. Fourth, hydrocarbon fluids must have found or created suitable pores in which to exist at depth and through which to travel in their journey upward, driven by buoyancy forces due to their low density compared with that of the rocks. Fifth and finally, a source of hydrocarbons must still exist at great depth. Can these five assumptions all be valid?

Five Assumptions Underlying the Deep-Earth Gas Theory

1. Hydrocarbons are primordial.

The first assumption underlying the abiogenic view of petroleum formation—that hydrocarbons were a common constituent of the primordial materials out of which the earth accreted—is now common knowledge among astronomers and planetary scientists whose domain of inquiry expands out to the breadth of this star system and beyond. But it must be remembered that the biogenic theory of petroleum formation was developed in the 1870s, before scientists had any notion that so-called "organic" molecules, including hydrocarbons, are in fact abundant in the universe. This fact of astronomy has been known since the early decades of the twentieth century, thanks to the invention of spectrographs that analyze wavelengths in the optical and radio portions of the electromagnetic spectrum. With these tools, chemical determinations have been made of distant bodies by capturing the spectral signatures of solar light either filtered through a planetary atmosphere or, less accurately, reflected by the surfaces of solid bodies that have no atmosphere. The consequences of these discoveries have not yet been fully integrated into the geological thinking of the present. The earth is, after all, a planet, and thus geology should be regarded first and foremost as a subset of planetary science, but that view has been slow to take hold. Because I spent a good part of my working life as an astronomer, I was made aware of the importance and the reliability of these observations early on.

What have the spectrographic studies revealed? Fundamentally, they have demonstrated that carbon is the fourth most abundant element in the universe and also in our solar system (after hydrogen, helium, and oxygen). Among planetary bodies, carbon is found mostly in compounds with hydrogen—hydrocarbons—which, at different temperatures and pressures, may be gaseous, liquid, or solid. Astronomical techniques have thus produced clear and indisputable evidence that hydrocarbons are major constituents of bodies great and small within our solar system (and beyond). The greatest quantity is found in the massive outer planets and their satellites. Jupiter, Saturn, Uranus, and Neptune have large admixtures of carbon in their extensive atmospheres, chiefly in the form of hydrocarbons—mainly methane. Titan, a moon of Saturn, has methane and ethane (CH_4 and C_2H_6) and several other hydrocarbon molecules in its atmosphere. Much like water in our own atmosphere, these hydrocarbon molecules are responsible for the clouds we see on Titan, presumably precipitating as rain onto methane–ethane lakes or seas. The temperature at that distance from the sun (9.5 times more distant than is the earth and thus receiving only a little more than 1 percent of the radiation intensity we receive here) puts these compounds just into the range where they can exist in liquid or vapor form, whereas water on the surface there could of course be present only as very cold ice.

Planets and their moons are not the only reservoirs of hydrocarbons in our star system. Many of the asteroids—the swarm of minor planetary bodies between Mars and Jupiter—also seem to have hydrocarbons on their surfaces and probably in their interiors. The recent flyby of a European instrumented spacecraft past Halley's Comet strongly suggests that hydrocarbons coat the surface of that icy body too. Indeed, all the planetary bodies of the solar system appear to have formed initially by the aggregation of solids.

Here at home we find further evidence that hydrocarbons were indeed a common constituent of the accreting earth. Meteorites colliding with the earth even today provide samples of the ancient materials from which planets formed. Those of the carbonaceous chondrite class contain some volatile substances, and it is widely held that this class supplied the earth with most of its complement of volatiles. Although carbon is a minor constituent of other types of meteorites, it is present at a level of several percent in the carbonaceous chondrites, mostly in unoxidized form, with a certain fraction in the form of hydrocarbon compounds. The Russian isotope investigator E. M. Galimov has made a strong case that the earth acquired a good proportion of carbonaceous chondrite material, because many isotope ratios of volatile elements in our atmosphere match closely the ratios in those meteorites but are substantially different from the ratios found in other meteoritic materials. Galimov shows the earth's isotopic ratios of two stable isotopes of hydrogen and carbon, as well as those of neon, argon, and xenon (three inert gases) to be similar to those in the carbonaceous chondrites, but substantially different from the values on other types of meteoritic materials; for neon the difference is as large as a factor of 500. By concentrating on the noble (chemically inert) gases in making these comparisons, Galimov assured himself that he would see samples that were not contaminated with the jumble of different materials of the crust.[5]

It is therefore clear that the occurrence of hydrocarbon molecules within the earth is in no way an anomaly. It would be surprising indeed if the earth had obtained its hydrocarbons only from a source that biology had taken from another carbon-bearing gas—carbon dioxide—which would have been collected from the atmosphere by photosynthesizing organisms for manufacture into carbohydrates and then somehow reworked by geology into hydrocarbons. All this, while the planetary bodies bereft of surface life would have received their hydrocarbon gifts by purely abiogenic causes. (Remember that carbon is an element, and no processes on earth—other than human-built nuclear reactors—can create it. I am sure there were no big stagnant swamps on Titan or Pluto.)

Conventional thinking within astronomy, along with meteorite studies, thus confirms the first key assumption underlying the abiogenic theory. Hydrocarbons—and unoxidized carbon—were important constituents of the materials from which the earth was assembled. Oxidized carbon was not. It is therefore a strange assumption to consider carbon dioxide as the primary carbon source that the earth provided for its nascent life. Among surface life, carbon dioxide is indeed the carbon source, but it does not necessarily follow that carbon was supplied to the atmosphere in this form, constantly and over most of geological time. The earth's highly oxygenated atmosphere would have ensured the transformation of upstreaming hydrocarbons into carbon dioxide soon after their emergence from the earth's crust. At deep levels in boreholes, hydrocarbons are much more abundant than carbon dioxide, an observation that confirms the deductions from astronomical and meteorite evidence.

2. The earth was subjected
to only a partial melt.

What about the second assumption, that the forces at work in a young aggregating earth and thenceforward for more than four billion years did not heat and rework the materials into a state of chemical equilibrium? Could a significant amount of those primordial hydrocarbons have remained unoxidized?

Until the middle of the twentieth century, it was thought that the earth had formed as a hot body, that it had been a ball of liquid rock—as such, no doubt well mixed—and that it had then gradually cooled, providing a differentiated crust overlying a homogeneous mantle. In such an evolutionary history, no primordial hydrocarbons could have survived the molten state. Even if hydrocarbons had initially been supplied in the formation process, then according to the molten-Earth viewpoint, they surely would have been destroyed soon thereafter. To account for the supply of excess carbon to the surface, advocates of this theory thought in terms of oxidized carbon only, because that would be the stable form expected in such a case. Carbon dioxide was indeed found to come out of volcanoes, which apparently confirmed this viewpoint.

It has become quite clear now that our planet, as well as the other inner planets and the satellites of the outer planets, all accreted as solid bodies from solids that had condensed from a gaseous planetary disk. The primary condensates, ranging in size from small grains to asteroid-size planetismals, all contributed to the formation of the final earth. In the early earth, partial melting did take place, causing melts of lower density to make their way to the surface while, presumably, melts of higher density sank down toward the center. The heat that generated this melting was the product of radioactivity contained in the material, as well as the heat resulting from gravitational compression. Once partial melting occurred, two other sources of heat came into play. For one, gravitational energy was released as materials moved and sorted themselves according to density. Second, there was the chemical energy of spontaneous reactions among mixing materials, because the original materials that were accreted as cold objects would not have been in the lowest chemical-energy configurations. The low-density partial melts produced the rocky layer that we call the crust. This crust covers nearly all the surface, and every basement rock could be seen to have once been a liquid magma or a partially molten aggregate, so scientists were left with the impression that the earth had frozen from an initial melt.[6]

This picture of a once-liquid earth was adopted, and it shaped much of the discussion in the early days of geology. Even though by now it is quite clear that only a partial melting was involved and that the bulk of the planet had never been molten, the thorough reevaluation of geological theory that such a change should have inspired has never occurred. Nowhere is this more evident than in the discussion of the origin of volatile substances on the surface: the water of the oceans, the nitrogen of the atmosphere, and the carbon-bearing fluids that appear to have been responsible for a great enrichment of the surface with carbon.

Within an initially molten earth, the volatiles would have come to the surface in the first phase. Later, when such a body had cooled, there would be little expectation of a renewed supply of volatiles from below. Quite the opposite, however, would be the expectation on a cold body that was heating up: Successive layers would reach temperatures at which volatiles would be driven off. Outgassing processes would be expected to continue as long as internal temperatures were increasing in any part of the body. Furthermore, there would be quite different expectations of the chemical nature of the various volatile substances. On a hot early earth, most fluids would be brought to the lowest chemical energy configuration early on, and later they could not provide any source of energy. In contrast, on a cold body that was heating up, the fluids that were produced would often be out of chemical equilibrium with their surroundings and could thus be a source of chemical energy.[7]

As noted in Chapter 2, a source of chemical energy from within the earth is a foundational premise for the deep hot biosphere theory. If there is no chemical energy to be exploited—that is, if all substances within the earth have come to chemical equilibrium—then the only energy source for earth life would be sunlight falling on the surface. An understanding of the oxidation state of carbon within the earth is thus of central importance.

The question of the stability of the earth's primordial supply of hydrocarbons against oxidation—that is, against combining with oxygen contained in the silicate and other minerals of rocks—is intimately connected with the details of the outgassing process. If the gases ascend in

regions of magma, then (as we have already discussed) chemical equilibrium between the hydrocarbons and the magma would be approached,
and this would usually favor oxidation of the hydrocarbon gases. Thus it
is no surprise that volcanoes generally emit carbon mainly in the form of
CO_2, with only minor amounts as methane, CH_4.

Where gases make their way through solid rock, however, the fate
of the hydrocarbons is altogether different. In that case, no chemical equilibrium between the rock and the gas need be expected. Many
investigators had based their considerations on such an equilibrium having been established, and that would preclude an ascent of
methane; it would all have been turned into carbon dioxide at deep
levels. Rather than churning in a brew of magma, encountering molecule upon molecule of potential oxidant, any gas upstreaming instead
through solids comes into contact with only a very limited amount of
rock on the surfaces of pores. A sufficient flow of hydrocarbons moving
through pores and cracks over a sufficient period of time will draw out
all available oxygen, thus allowing hydrocarbons later following the
same pathways upward to pass without compromise. What this means
is that although active volcanic features are the most obvious places to
sample gases that have risen from great depths, because of oxidation
processes they are the worst places to obtain a representative reading of
the composition of gases and other fluids at depth. The best places to
get a representative reading are areas removed from volcanoes and
from any other indications of magma dwelling beneath—just average
areas of ocean floor and continental surface.

Deep-earth gas theory thus depends, in part, on the validity of a
second assumption. Hydrocarbons must not only have been primordial
constituents of the newly accreted earth; they must also not subsequently have been fully oxidized. The earth must have been subjected
to only a partial melt.

3. Hydrocarbons are stable at great depth.

Now we turn to the third assumption on which the abiogenic theory of
hydrocarbon formation depends, the thermodynamic stability of

hydrocarbons at great depth. It used to be thought that temperatures above 600°C would dissociate even the simplest and most heat-resistant hydrocarbon, methane (CH_4), and that temperatures as low as 300°C were sufficient to destroy most of the heavy hydrocarbon components of natural petroleum. Because such temperatures are reached at depths of only a few tens of kilometers in the crust, it seemed pointless to discuss an origin of hydrocarbons from non-biological sources at deeper levels. If the origin had to be found in the upper and cooler parts of the crust, then there was really no alternative to the biogenic theory.

This conventional view on the thermal instability of hydrocarbons reigned unchallenged simply because the cost of conducting experiments at the appropriate pressures was prohibitive and the importance of doing so was not appreciated. Calculations of thermal stability that were undertaken in the West did not take into account the substantial effects of pressure: High pressure greatly stabilizes hydrocarbons against thermal dissociation. We must therefore assess the question of hydrocarbon stability not only at the temperatures, but also at the pressures, that prevail at various depths.

Thermodynamic calculations made by the geoscientist E. B. Chekaliuk and published in a Russian journal in 1980 indicate that methane would resist complete dissociation down to a depth of 300 kilometers, except in volcanic regions that breach the normal temperature gradient of the earth.[8] Perhaps a depth of somewhere around 600 kilometers would be the lower limit for the possible existence of methane within the earth.[9] (See Figure 3.1.)

What about the heavier hydrocarbon molecules that make up the bulk of petroleum? Thermodynamic calculations done in Russia and in the Ukraine have suggested not only that most of these molecules are stable in the pressure–temperature regimes that prevail at depths between 30 and 300 kilometers but also that they would be *generated* if a mix of simple carbon and hydrogen atoms were present at those depths.

At a depth of, say, 200 kilometers, a mix of hydrocarbon molecules would be the expected equilibrium configuration—and this despite

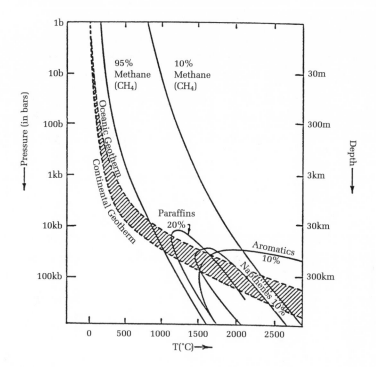

Figure 3.1 Stability of hydrocarbons at temperatures and pressures in the earth (from Chekaliuk, 1976). The vertical scale on the left represents pressure marked in bars, where 1 bar is equal to the pressure exerted by the atmosphere; thus the top of the diagram denotes the surface of the earth. The vertical scale on the right represents the depth corresponding to the pressure, assuming a mean density of rock of 3.5 times the density of water. The increase of temperature with depth in the earth is referred to as the "geotherm," and the region between the two geotherms is the region that might represent the temperature–depth relation in different locations. The deep ground under the oceans is generally hotter than deep ground at the same depth on the continents, as can be seen in the figure. Methane (CH_4) is the most stable molecule of the hydrocarbons; most of it would survive at all levels down to 300 kilometers, provided the temperature there did not exceed 2000°C. For the other components of natural petroleum—paraffins, aromatics, and naphthenes—the percentages in equilibrium are shown; these would be the values most likely to be produced from a mixture of hydrogen and carbon. Methane streaming from great depth could bring up, in solution, significant fractions of these petroleum components.

thermal conditions that, were pressure not taken into account, would be far in excess of the threshold for dissociation of these molecules. The detailed chemistry of the resulting molecules would depend on pressure, temperature, and the carbon–hydrogen ratio. Other atoms that might also be present, such as oxygen and nitrogen, would form a variety of complex molecules with the carbon and hydrogen.

It is very interesting to consider that complex molecules made of carbon, hydrogen, oxygen, and nitrogen are also made by life—but life can perform this feat even at the low pressures that prevail at or near the earth's surface. These molecules we call proteins. At the very great pressures prevailing at depth, a further degree of molecular complexification might also occur. Metal atoms present at such depths would combine with the hydrocarbons to form organometallic compounds (a prospect we shall examine in Chapter 7, because it may well bear on the formation of near-surface deposits of concentrated metals). Enzymes, catalysts for biochemical reactions, are often composed of metal atoms in complex molecules made with carbon, hydrogen, oxygen, and nitrogen.

4. Rock at depth contains pores.

What about the simple physical assertion that a vast amount of hydrocarbons can indeed remain at great depth within the earth's crust because pore spaces do in fact exist in those realms to accommodate their presence, and mechanisms do exist to facilitate their flow?

I first became interested in the question of whether pores exist at great depths during the early 1950s when I was still at Cambridge. What sparked my interest was a clearly erroneous statement I had come upon in a geology textbook, which I was reading more out of curiosity than with any particular question in mind. There it was stated that rocks porous enough to hold and transmit fluids must be restricted to a thin outer layer of the crust, not much deeper than the depths to which the deepest petroleum wells penetrate. Below that, the textbook explained, the weight of the overburden would be so great that even the strongest rocks would be crushed to a degree that all pore spaces would be eliminated; no fluids could be contained there, and no movement of any fluids through all the deeper rocks would be possible. A calculation was given, showing the measured crushing strengths of rocks and comparing them with the pressures that would be exerted by the weight of the overburden. Superficially this may have sounded convincing, and the numbers used were quite correct. What was wrong was the implication that fluids could not be contained at deeper levels, even if they were at a pressure similar to that of the rocks.

I remember discussing this with my friend and colleague at Cambridge, the astronomer Fred Hoyle. Jokingly I said, "This is about as silly as the question of a schoolboy who first learns about atmospheric pressure and asks why he is not squashed as flat as a pancake if there is a pressure of 14.7 pounds per square inch on his body." We discussed the situation of a "pressure bath," where everything—rocks and liquids and gases—is immersed at each level in a common pressure. Under such conditions there would be just as much porosity and permeability, in the form of connected pore spaces that allow fluids to migrate, as exist in near-surface rocks and sediments at low pressures. Just as the schoolboy was not squashed flat, so the deep pores would not be squashed out.

Recall from Chapter 2 that life along the deep-ocean vents must grow from tiny eggs into mature clams and tube worms at pressures 20 to 200 times greater those than we surface creatures experience. Living creatures are just as delicate down there as they are up here, and yet the high pressure of the deep ocean does not pose a problem for biological growth, which can be thought of as the creation of more and more "pores" in the form of cells. Pores within rock, like cells within living organisms, can be maintained at very high pressures, so long as the fluid that occupies the pore or cell exerts an outward pressure as great as the opposing pressure of the surroundings. It is this pressure balancing or differential, not absolute pressure, that determines the fate of a pore.

This argument in favor of deep crustal porosity first appeared in Hoyle's book *Frontiers in Astronomy,* published in 1955. Hoyle had written it up as a chapter entitled "Gold's Pore Theory." When I took up the subject again a quarter-century later, I had quite forgotten about that early discussion and Hoyle's publication of the idea, but I developed exactly the same notion. Only later did I find the chapter in Hoyle's book showing that I had said the same thing long ago!

Beginning at the earth's surface, rocks of the same composition do indeed tend to become less porous at depth, down to a level devoid of substantial connections between pores. At that level there would be great resistance to the flow of a liquid. Petroleum geologists therefore extrapolated from this curve of diminishing porosity to the conclusion that at even greater depths there would be even less porosity and permeability. But quite the opposite is true. At a critical depth limit, where the flow is greatly impeded by the compression of the rock, this incomplete but nevertheless effective barrier allows a higher pore pressure to be built up behind it by fluids under higher pressure coming up from below. In this somewhat deeper domain, we may therefore expect to see again a high porosity and permeability of the rock. Instead of having the idealized "pressure bath" discussed earlier, a rock that has a finite compressive strength will set up a stepwise approximation to it. When the crushing pressure on the rock is reached, the rock will compress to the low-permeability state. (Crushing pressure is determined by the overburden weight of rock less the fluid pressure given by the head of fluid, which is usually only about one-third of the former because most fluids are about one-third as dense as the rock.) At yet deeper levels the same pattern may occur again, and there may be several cycles before the rock is so hot that it lacks mechanical strength, at which stage the ideal pressure bath will be established.

Conventional criticism that hydrocarbon fluids cannot exist at depth for lack of pore space thus does not stand up to scrutiny.

5. Hydrocarbons are still upwelling.

The final assumption on which the abiogenic theory of petroleum formation depends is that to account for the hydrocarbons now available in the crust, a hydrocarbon source from which those fluids upwelled is still present at great depth. A variety of empirical evidence at or near the surface argues in this direction. In Chapter 4, we shall see that the *chemical* data (carbon isotopes) that critics present as the strongest refutation of the abiogenic theory actually argue in its favor. In Chapter 5, I will show how the *biochemical* data can be seen as consistent with the abiogenic theory. In Chapter 6, I will present a very strong challenge to the biogenic theory by relating the results of deep drilling in a geological province (entirely igneous) where, according to conventional theory, hydrocarbons simply cannot exist.

Chapter 4   Evidence for Deep-Earth Gas

T he abiogenic theory of petroleum formation presumes that an enormous source of primordial hydrocarbons resides in the upper mantle and lower crust—far deeper than can be drilled and sampled directly. Consequently, we must search for evidence in the regions of crust that our drilling equipment can penetrate. What empirical evidence in the shallow crust or at the earth's surface favors the theory that a substantial source of hydrocarbons exists at great depth?

The empirical evidence is of seven main types.[1] First, reservoirs of petroleum, including the various gaseous forms such as methane and ethane, are frequently found in geographical patterns of long lines or arcs extending for hundreds or even thousands of kilometers. The island-studded arc of Indonesia is perhaps the best example. These linear patterns are related more to deep-seated and large-scale structural features of the crust than to the smaller-scale patchwork of the sedimentary deposits. Dmitry Mendeleyev, the Russian chemist who originated the periodic table of elements, noted these large-scale patterns of the occurrence of hydrocarbons in the 1870s, and much new information has greatly strengthened the case.

Second, petroleum deposits that have been discovered follow what is known as Koudryavtsev's rule: Hydrocarbon-rich areas tend to be hydrocarbon-rich at all lower levels, corresponding to quite different geological epochs, and extending down to the crystalline basement that underlies the sediment.[2] Russia's great petroleum geologist of the early part of the twentieth century, N. A. Koudryavtsev, cited many examples from all over the world that clearly showed this depth effect, as many subsequent Russian petroleum geologists have also done. Even where drilling has penetrated past the sedimentary strata and into the basement rock, evidence of hydrocarbons does not run out. Invasion of an area by hydrocarbon fluids from below could better account for the vertical reach of hydrocarbons than does the chance of successive deposition of hydrocarbon-producing biological sediments in epochs that differ by tens of millions of years and that show no similarities of climate, vegetation, or other relevant characteristics.

Third, methane is found in many locations where a biogenic explanation for its presence is improbable or where biological deposits seem inadequate to account for the size and extent of the methane resource. These anomalous locations include the great ocean rifts (which lack any substantial sediments); fissures within rocks that had clearly frozen from a melt at a temperature too high and a pressure too low for any pre-existing hydrocarbons or biological remains to have persisted in such extreme conditions; and depths far below any sediments that contain biological materials. (Chapter 6 will provide a detailed look at one such occurrence of hydrocarbons in Sweden.) In addition, as noted in Chapter 2, huge amounts of methane hydrates have been discovered covering vast areas of the ocean floor. Methane hydrates are also present in large amounts in permafrost ground.[3] Their widespread distribution indicates that many or most regions of the crust emit some methane—enough over long periods of time to saturate any domain in which this ice is stable. The outflow rates may be quite variable regionally, however. The view that the main carbon supply to the surface comes from volcanic emission of carbon dioxide is thus in doubt, so long as no estimate has been made of the sum of the diffuse outflows of methane over all sea and land surfaces.

Fourth, the hydrocarbon deposits of a large area often show common chemical features regardless of the varied composition or the geological ages of the formations in which they are found. Such chemical "signatures" may be seen, for example, in the abundance ratios of some minor constituents, such as traces of certain metals that are carried in petroleum (the most common but by no means the only ones are nickel and vanadium), and in a shared tendency to favor some of the different molecules that make up petroleum. Thus a chemical analysis of a sample of petroleum often makes it possible to identify the general area of its origin, even though the oils of that region may be coming from a wide variety of geological formations.

Fifth, as many observers have noted, a number of hydrocarbon reservoirs seem to be refilling as they are exploited for commercial production. As I shall soon explain, the abiogenic theory would account for this observation. I do not think the biogenic theory could—at least, I have not heard of any prediction of the refilling phenomenon. It has simply been observed and noted.

Sixth, the distribution of the large amounts of carbonate rock in the upper crust and the isotopic composition of the carbon atoms within it argue against the theory of a surface biological origin of most of the buried hydrocarbons.

Seventh and finally, the clear, well-established regional associations of hydrocarbons with the chemically inert gaseous element helium have no explanation in the theories of a biological origin of petroleum. But as we shall see, these associations are explained if the hydrocarbons have ascended from great depth.

Of these seven classes of empirical evidence favoring the abiogenic theory of petroleum formation, the first four are well known in the petroleum business. The brief descriptions just provided are adequate for our purposes, and I shall not expand further on them here. For the last three points, however, I can offer some arguments that were previously not known or were not adequately taken into account.

Petroleum Reservoirs That Refill

When a new oil or gas field is explored, an observation is routinely made of a drop in pressure resulting from a given volume of production. Measurement of this change is used to estimate the total volume accessible to the wellbore. Aggregated worldwide, these estimates of reserves drive petroleum exploration and, to some extent, the economic outlook for industrial nations. But it has turned out that such estimates are nearly always much lower than actual production over the course of many years. The same error in the initial estimates was also the reason for the belief, widely publicized in the early 1970s, that the global supply of crude oil would be exhausted within fifteen years. This dire prediction profoundly affected the price of petroleum—and through that, the distribution of wealth among nations.

Under the abiogenic theory, if oil and gas are flowing upward from deep (and thus high-pressure) levels, their travels cannot be arrested by any caprock, however competent the rock may be. No rock has a significant tensile strength, so no rock can hold down a fluid that comes up with a pressure greater than that exerted by the weight of the overburden. A caprock will create a concentration of the fluids below it, but the steady flow rate will eventually be reestablished at a value equal to the flow rate at the deep source. The flow through a caprock obstruction is thus like that of a river crossed by a dam. The dam causes a lake to form on the upstream side, but after the lake has filled, the flow rate resumes. The same amount of water will flow over the dam as the river carried before the dam was built.

One might think that the upwelling flow of hydrocarbons could itself provide the recharging mechanism responsible for reservoir refilling. The upwelling flow will do so to some degree. But if the upflow were as fast as the recharging observations indicate, then the rate at which carbon is delivered into the atmosphere would be much higher than atmospheric observations allow. There is, however, another process that can cause a much faster refilling without driving more carbon up into the atmosphere.

As already noted, rock that contains a fluid of lesser density in its pores will inevitably set up a pore pressure regime in which the fluid pressure defines stacked domains, each separated from the one below by a layer of crushed rock of very low porosity and permeability. If oil and gas have indeed come up from below, we can expect a vertical series of deeper reservoirs to be stacked below the producing field. If, now, the uppermost domain has its fluid pressure decreased by production of oil

or gas, then the pressure differential across the crushed layer of low permeability will automatically increase. Transport of fluid through that
layer will therefore accelerate. The top field will be replenished at a rate
given by the leakage from below, when the delicate pressure balance
between rock and fluid has been changed. The top field will be drawing
on the deeper reserves that have not been accessed directly. In the course
of time, at a slow rate given by creep deformation in the rock, the stepwise pressure pattern will adjust its levels to the new pressure situation.
In other words, without drilling any deeper, we can nevertheless tap into
the deeper reserves that may well be much larger than the reservoir under
production. The mean rate of outflow from the deepest source of the
hydrocarbons will not have increased; rather, more of the fluids that
already exist at intermediate depths will have become accessible.

The phenomenon of petroleum reservoirs that seem to refill themselves is widely reported, notably in the Middle East and along the U.S.
Gulf Coast.[4] I regard these occurrences as strong evidence for the deep-
earth gas theory.

Clues in the Carbonate Record

The surface and subsurface sediments on the
earth contain approximately one hundred
times as much of the element carbon as
would have been derived from the grinding up of the basement rocks
that contributed to the sediments. The surface is thus enormously
enriched in carbon. This enrichment requires an explanation.

The total quantity of carbon contained in the sediments and on the
surface is estimated to average about 200 tons for each square meter of the
earth's surface area. One-fifth of all this carbon is in *unoxidized* form,
including various grades of coal, crude oils, kerogen (carbonaceous compounds diffusely distributed in the rocks), and natural gas, either as free
gas or in the form of methane hydrate ices. In addition there is the thin
veneer of living and not-yet-decomposed biological material. This latter
category—in my opinion the only demonstrably biological component—
represents only a very small fraction of the total unoxidized carbon.

The other four-fifths of the carbon is in *oxidized* form, mostly lime-stone (calcium carbonate, or $CaCO_3$) and dolomite (a blend of calcium and magnesium carbonate).[5] Much of this carbonate was deposited in oceans, having derived the carbon from the atmospheric–oceanic pool of CO_2. Carbonate precipitates naturally out of the water column from dissolved carbon dioxide and calcium or magnesium oxides. It can also be precipitated out of the water biologically, by organisms that build carbonate shells or skeletons.

One attempt at an explanation of this large excess of carbon at the surface and in the sediments was to suppose that in the early days of planetary accretion, the earth acquired a huge atmosphere of carbon dioxide, which was then turned into carbonate rocks. Later, subduction of some of the carbonates carried along the ocean floor and into the plunging boundaries of tectonic plates would transport the rocks to depths at which the carbonates would dissociate. Carbon dioxide would be released, and it would be returned to the atmosphere in vol-canic eruptions. A fairly steady rate of this carbon flow, cycling between deposition as carbonate and release as carbon dioxide, was proposed to account for a continuous supply of atmospheric carbon dioxide over geological time, at least over the last well-documented two billion years. (Earlier times do not offer useful data, except to indi-cate that some carbonate rocks did indeed exist more than three billion years ago.)

According to this explanation of the earth's near-surface enrich-ment in carbon, the initial blanket of carbon dioxide in the earth's atmosphere would have to have been very substantial. The figure implied by the mass of carbonate rock mentioned would require a mass of carbon dioxide in the early atmosphere eighty times greater than the whole of our present atmosphere and about as massive as that of our sister planet, Venus. In contrast, today's proportion of carbon dioxide in the earth's atmosphere is only 3.5 parts per ten thousand, by volume.

However, there is good reason to believe that the early earth did not acquire much material in the form of gases, because there is a very low abundance of gases such as neon, non-radiogenic argon, krypton, and xenon in the atmosphere today. No physical process could have sorted out these inert gases from the solar system's gaseous mix, where they are known to be considerably more abundant. And because all these inert gases are heavy atoms, they would not have escaped the earth's gravity and drifted off into space at a greater rate than other gaseous elements. The only sound explanation, in my view, is that atmospheric gases have derived mainly from outgassing of volatiles derived at depth from buried solid materials—not from an initial large atmosphere acquired at the earth's formation or by later capture of gases from space.

The theory that the earth started out with a massive CO_2 atmosphere fails in yet another way. The pattern of carbonate rock deposition through geological time does not support it. Rather than a skewing of carbonate deposition to earlier times, the sedimentary record shows a rather continuous accumulation of such oxidized carbon, as well as unoxidized carbon, over the last two billion years—which is the period of time over which the sedimentary record is usefully intact. Indeed, the total carbon excess of the surface layers is clearly shown to have been increasing since early times. Recycling cannot account for that. Rather, a continuous addition drawn from sources upwelling from within the earth must be held responsible.

Strangely, although most of the oxidized carbon that is in the carbonate deposits is derived from the atmospheric–oceanic pool of carbon dioxide, the present content of carbon in this pool represents only about one part in 740 of the known deposited amounts (using the estimated total deposited carbon over the course of two billion years and the measured CO_2 content of atmosphere and oceans). What is the origin of the supply that maintains atmospheric CO_2 at levels that result in the deposition of carbonates through all geological epochs and that maintains a supply rate sufficiently constant to keep plants alive?

If outgassing of carbon-containing volatiles from the depths of the earth were responsible, what mean rate of outflow would be implied? Using the figures presented above, this global average rate of outgassing would have to be sufficient to replace the amount equal to the present atmospheric–oceanic content of carbon dioxide every 2.7 million years. In other words, the carbon must have been replaced in those surface reservoirs 740 times in two billion years.

As already mentioned, the chemistry of meteorites indicates that carbonates or other forms of oxidized carbon were not common constituents of the materials that formed the solid planets. Most of the carbon was initially in unoxidized form, primarily as hydrocarbons. The evidence from deep boreholes that are not too close to active volcanic regions shows, in accordance with the meteorite evidence, that hydrocarbons are the dominant carbon-bearing fluids there. At still deeper levels, where the pressure is so great that diamonds are the stable form of carbon, unoxidized carbon again evidently dominates and forms these crystals of pure carbon. (Chapter 7 will further explain.)

Some fraction of these upwelling carbon fluids, starting out largely in the form of CH_4 and other light hydrocarbon molecules, will be oxidized during the ascent. The oxygen availability from the rocks, the temperature and pressure along the pathways of flow, and the action of subsurface microbial life will determine the ratio of methane to carbon dioxide emerging from the ground in any one region. Any methane that reaches the atmosphere without being oxidized along the way would quickly be oxidized to carbon dioxide in the oxygen-rich atmosphere and there join the pool of atmospheric–oceanic CO_2. What fraction of all the upwelling carbon volatiles would be delivered to the atmosphere as methane, and what fraction as carbon dioxide? The carbon dioxide coming from volcanoes is well studied, whereas the large quantities of methane that emerge from non-volcanic ground go mostly unnoticed. The (superficial) impression created by this is that carbon dioxide is the principal source of the surface carbon excess, and that it is also the main carbon-bearing gas in the ground.

An analysis of the isotopes of carbon, however, reveals an error in this dominant view. The study of the isotopes of carbon is a large and complex field. I will mention here only one aspect that bears directly on the subject under discussion, but even that is necessarily rather technical. It has to be addressed because there has been much debate about its interpretation and significance. (Readers not inclined to absorb this level of detail could skip to page 68.)

Natural carbon has two stable isotopes: carbon-12 (C-12), which has 6 protons and 6 neutrons, and C-13, which has 6 protons and 7 neutrons. In earth materials, about one in a hundred carbon atoms is the heavy isotope, C-13; the rest are C-12. No chemical action can change one of these stable isotopes into the other. The ratio that is seen is inherited from the nuclear processes in stars that assembled this carbon. All that can happen on and within the earth (leaving aside nuclear reactors) are selection processes—isotopic fractionation—that favor the movement of one or the other isotope. The chemical reactions of the two isotopes are closely similar, and no significant chemical fractionation can be expected. The most significant difference between them that could cause fractionation is the difference in their masses, which brings about a difference in the velocity of their thermal motion. This will cause the two molecules to move at different speeds in circumstances in which the flow speed is influenced by the thermal motion speed, such as in molecular diffusion through a finely porous material. In such flows, a marked fractionation can be expected in many cases.

For a light molecule such as methane, isotopic fractionation would be an important effect. Methane has a molecular mass of 16 units when made from the light isotope of carbon, a mass of 17 units when made from the heavy. Diffusion speeds would be 3 percent faster for the light molecules. Thus if a stream of methane were to flow over a semipermeable membrane, we might expect that on the other side of this membrane the proportion of light methane would be enhanced, quite possibly by 3 percent.

If the carbon isotopes were contained in molecules of carbon dioxide rather than methane, isotopic fractionation would still occur, but it would be less pronounced—only 1 percent—because the molecular mass units would be 44 against 45 for the two isotopes of carbon. The proportions of the two carbon isotopes in carbon dioxide sampled in the air or water are virtually the same the world over. This has been taken to mean that the fluid reservoirs on the earth's surface are supplied with carbon dioxide from a single source of that particular isotopic ratio. But this does not follow at all. Any carbon gas that enters the atmosphere is globally mixed in a short time compared with the time taken for it to be fixed into a solid. It is therefore just the global average of the isotopic contributions from the two carbon gases that this single measured value represents. Both methane and CO_2 may have contributed, with varying isotopic ratios.

To understand the importance of carbon isotope data for the deep-earth gas theory, we must now turn to the isotopic ratios. All later stages in the food chain of the surface biosphere carry through the same ratio of light to heavy carbon as appears in the photosynthesizers. Plants and algae are invariably deficient in C-13, compared with the proportion maintained in the atmosphere. Photosynthesizers select in favor of the lighter isotope, C-12. When this fact was first discovered, the process by which life fractionated carbon isotopes was unknown. Because photosynthesizers everywhere performed the same trick of fractionation, the opinion developed that life alone was responsible for C-13 deficiencies anywhere in the rock record.

We now know that the way life achieves this isotopic fractionation is through passive diffusion, not active mediation and control. The lighter C-12 passes more readily through the pores of semipermeable membranes that are a part of the equipment of photosynthesizers. Thus *any* process by which a carbon-bearing molecule passively diffuses through a porous mass—whether that mass be living substance or non-living—should result in fractionation. Might geological processes produce such fractionation?

Yes. Upwelling methane would be subject to fractionation wherever it passes through a wet spot in the rocks or a particularly tight network of pores. Indeed, very large factors of fractionation can be obtained by multiply stacked diffusion systems, which may easily be encountered during the upward journey of hydrocarbons originating at great depth. Such an extended process of fractionation would account for the extreme values of C-13 deficiencies recorded for hydrocarbons sampled in some locations—greater values than have been reported in plants anywhere. Nevertheless, the interpretation that only biology could produce significant fractionation has been adopted so overwhelmingly that *whenever* the light carbon isotope is found to be favored in a subsurface solid or fluid, life is unquestioningly held responsible.[6]

Geochemical analysis of samples of natural gas drawn up from the earth's crust everywhere in the world shows that over 99 percent of the hydrocarbon gases are enriched in the lighter isotope of carbon, compared with the pool of carbon isotopes contained in the atmosphere and oceans, although the spread in enrichment values is rather wide. This fact has been used to support the biogenic theory of the formation of hydrocarbons. But the abiogenic theory would also offer a satisfactory explanation. Methane molecules bearing the heavier isotope of carbon would ascend more slowly through the rock than would molecules bearing the lighter isotope. Diffusion from pore to pore would be slowed. The longer transit time would subject those heavier molecules to more opportunities for oxidation. The carbon component of many of those heavy methane molecules would thus oxidize into carbon dioxide, which would continue the ascent. Carbon dioxide emanating from the earth is, in fact, enriched in the heavy isotope compared with hydrocarbons. To test this explanation, however, would require far more accurate data—not only on the isotopic differences between methane and carbon dioxide outgassing but also on the relative quantities of each released into the atmosphere. Here the ratio of methane to carbon dioxide outgassing would become important.

Both the biogenic theory and the abiogenic theory of the formation of hydrocarbons can thus explain the isotopic signature of hydrocarbons found within the earth's crust. But I maintain that only the abiogenic theory can satisfactorily account for the carbon isotope composition of the carbonates that constitute the major component of the earth's inventory of surface carbon. To grapple with this issue, a bit of background is first required.

In the earth sciences, precise measurements of small variations in the carbon isotope ratio are used to make deductions about the history of carbon materials bound in the rocks. These measurements are generally reported as the deficit (negative) or the excess (positive) of C-13, relative to a standard carbonate rock selected as a typical marine carbonate. The results are generally plotted as departures from this standard, in parts per thousand. It is also important to know that carbonate rock that is laid down in the oceans bears about the same carbon isotope signature as the atmospheric–oceanic pool of CO_2 at the time of its deposition. Little fractionation occurs during physical precipitation of carbonates. Little fractionation occurs even when a limestone is built largely from the calcium carbonate shells and skeletons manufactured by life. This is because living organisms precipitate carbonate from the surrounding water for their own uses; they do not extrude carbonate through cell membranes. Thus there is no diffusion involved.

Under the biogenic theory, the entire inventory of crude oil, natural gas, coal, and kerogen, as well as methane hydrates, would constitute a removal of carbon from atmospheric CO_2. Such deposits have increased over geological time and constitute an ongoing process. The carbon laid down by vegetation would be deficient in C-13 compared with the atmospheric CO_2 from which it was thought to have derived, and a cumulative shift in favor of the heavy isotope of this atmospheric CO_2 would result. Because the oceanic carbonate rocks got their carbon from this CO_2, the carbonate record should show a gradual increase in the proportion of C-13. Given the quantities estimated for the deposits and their isotopic ratios, this effect should be sufficiently large to be observed in carbonates laid down over geological time.

But no such effect is seen. The carbonate deposits in fact show a small range of the isotopic ratio, which has stayed remarkably constant from early Archean times to the present.[7] The biogenic theory fails to account for this fact. This imbalance could not be redressed by the recycling of the unoxidized carbon deposits; these would largely turn into insoluble and heat-resistant elemental carbon.

On the basis of the abiogenic theory we would consider most of the unoxidized carbon deposits in the crust as derived from upwelling hydrocarbons, not from any sediments coming from the atmosphere. Unoxidized carbon deposits could therefore have no effect on the isotope ratio of atmospheric CO_2. The fractionated carbon of the plants would all be returned to the atmosphere in the decomposition process of surface life, so fractionation by living cells would not affect the average isotopic composition of the atmosphere.

The form in which carbon is delivered to the atmosphere, oxidized or unoxidized, also has an important effect on the quantity of oxygen available to the atmosphere.[8]

Another class of carbonates exists—one that certainly did not derive from atmospheric carbon dioxide. These are the crack-filling carbonates, therefore called cements, that are found in many rocks. Carbonate cements are common in petroleum-bearing areas. In fact, by making the rock above a reservoir less permeable, they may play a significant role in retaining hydrocarbon pools large enough for commercial exploitation.

Three independent attributes set carbonate cements apart from the bulk of marine carbonates. First, they are crack-filling and not layered. Second, they are more abundant in petroleum-bearing areas. Third, they show a far greater spread of the isotopic ratio than do any other carbonates (Figures 4.1 and 4.2).

I attribute the wide isotopic spread of the carbon in carbonate cements to the spread in the isotopic ratios of the upstreaming methane from which the cements derive. Methane, because of the relatively great difference in mass between its two isotopic forms, would be expected to suffer much isotopic selection as it flows through water or tight spots in the rocks. Some of the upstreaming methane oxidizes in the rocks. The carbon dioxide so created then reacts readily with calcium oxide present in the same rocks, forming carbonate. That carbonate reflects the isotopic variant that the methane brought in.

Whenever I discuss the large scatter of the isotopic ratio seen in the carbonate cements and offer my explanation drawn from the deep-earth gas theory, I can expect to be confronted with statements such as "You are not aware that this must be due to a mixture of two different sources of methane in all the regions you investigated." But there are hundreds of examples drawn from core samples of oil wells, from carbonate deposits on the ocean floor that overlie gas production areas, and of course from our wells in Sweden. Would they all have acquired gases in comparable amounts from two different sources, one of them being biogenic? Would there not be many locations where only one source and not another had contributed? How would biogenic methane have made its way down 500 meters (or indeed, six kilometers) into the granite of Sweden, starting from a surface that had a sedimentary cover barely sufficient to hide the bedrock?

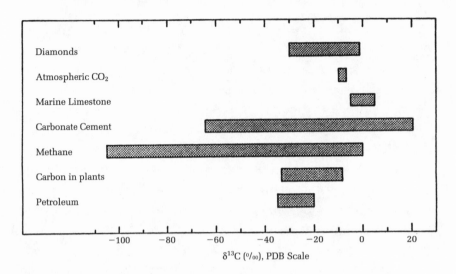

Figure 4.1   The wide distribution of carbon isotopic ratios in methane from all natural sources. Note that the range of atmospheric CO_2 is small, as it must be because all diverse contributions are quickly mixed globally in the atmosphere. The oceanic carbonates that derived their carbon from the atmospheric CO_2 show a similarly small range. SOURCE: Thomas Gold, 1987, *Power from the Earth* (original compilation from values given in various textbooks).

Geological fractionation of methane obtained through a difference of diffusion speeds is not an outlandish speculation on my part. Such a process is employed on a large scale in industry, for example, in the processing of uranium for use in bombs or nuclear reactors. In industrial processes, fractionation is often used in many successive stages, and very large fractionation effects can thus be obtained. In the less well organized circumstances of porous rocks, the same fractionation effects take place with lower efficiency.

The transport of a gas through rock happens in two types of flow. First, there is a bulk stream that flows through connected pathways created and held open by the gases derived from deeper and higher-pressure levels. The flow speed in the bulk stream is shared by all the molecules of gas, regardless of isotopic content. That flow speed is determined by pressure gradients and viscous friction within the rock.

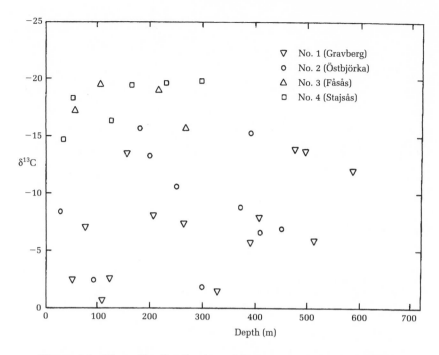

Figure 4.2 The wide distribution of the isotopic ratios in carbonate cements in four shallow (less than 600 meters deep) boreholes in the impact formation of the Siljan Ring in central Sweden. Note the large changes seen over distances of a few tens of meters only. Gravberg is in the area of the first deep drilling that was carried out in the Siljan Ring. In contrast, marine-layered carbonates would almost all lie on the vertical axis between −4 and +4.

The second type of flow is due to diffusion of the gas into a multitude of capillary pore spaces, often water-filled. There, speed is given by the individual motions of the molecules, and there fractionation will occur. The methane that reaches the atmosphere from deep levels and through long pathways, by which it will have suffered diverse amounts but large values of fractionation, emerges with all the signs of this fractionation removed as soon as atmospheric mixing blends it into the worldwide reservoir. The constancy of the atmospheric ratio of carbon isotopes everywhere and through time thus does not mean that fractionation of hydrocarbons was absent in all the individual outflow areas; it means only that the *averages* for the entire earth and for periods of millions of years were closely similar.

The isotopic evidence must be considered in any discussion of carbonate genesis. The only explanation that will avert the conclusion that the proportion of the heavy isotope in marine deposits of carbonates must increase through time would be that most of the deposits of unoxidized carbon in the ground (oil, gas, coal, and hydrates) did *not* derive from the debris of photosynthetic surface life. Rather, they must have derived from unoxidized carbon that came up from depth. This explanation will be coupled with the conclusion that outgassing of the primordial complement of carbon, mainly in the reduced form of methane, has occurred at a reasonably steady pace. Any methane that makes its way up through the rocks and to the surface would in any case be oxidized in the atmosphere to carbon dioxide, and we could no longer distinguish it from gas that had entered the atmosphere already in fully oxidized form.

In sum, the technical information and arguments in this section lead, in my view, to a straightforward general conclusion: The volumes, ages, and isotope ratios of crustal carbonates represent important evidence in favor of the view that hydrocarbons were primordial constituents of the earth, that they remain still, and that they continuously upwell into the outer crust, finally emerging, oxidizing, and mixing in the atmosphere.

The Association of Helium with Hydrocarbons

There is a very strong association of helium with hydrocarbons. This association is so strong that in the commercial search for hydrocarbons, helium sniffing along the surface has been found useful. Very sensitive helium detectors now exist. They were first employed for the detection of uranium deposits underground, which were thought to be major sources of helium. That search was not successful. But helium sniffing did prove helpful in detecting oil and gas fields.

The association of helium with hydrocarbons is probably the most striking fact that the biogenic theory fails to account for, and therefore it has been for me of the greatest interest.

Where does that helium come from, and why is it so strongly associated with hydrocarbons? In the rocks, helium is produced mainly in the radioactive decay of uranium and thorium. Regional patterns of helium abundance have been observed in which the present quantities of helium are far higher than the sediments could ever have produced from the total of their radioactive components. In these regions of helium abundance, the mixture of gases (including hydrocarbons and nitrogen) within and emerging from the earth's crust tends to be remarkably similar over very large geographical areas, spanning even very different geological provinces. The helium therefore must certainly have come from below the layers of sedimentary rock, and it must have arrived there already in regionally well-defined mixing ratios with methane and nitrogen, so that the different fields of the region could all be filled with the same or a closely similar mix. Only a mix that had entered the sediment and its individual gas fields from below could achieve that effect.

Any chemical or biological cause of this enrichment can be ruled out for the chemically inert helium, which does not establish chemical bonds with any element. No chemical process—biological or nonbiological—can cause helium to be gathered up from a low concentration and brought to a higher one. Only variations in the concentrations of the parent radioactive elements, and variations in the length of the pathway through the rocks from which the helium has been swept, could explain the great regional differences in observed helium concentrations. Where helium concentrations have varied widely from one location to another, such as by factors of a hundred or more, the length of the pathway through which the carrier gas has swept is likely to have been the dominant variant.

Beginning in about 1979, I began working with the vast accumulation of helium data already available in order to find a reasonable explanation for the regional patterns. Perhaps the helium enrichment data could serve as a good proxy for the depth from which its fluid hydrocarbon carrier began the upward journey. The deeper the source of hydrocarbons, the greater the total length of pore spaces through which hydrocarbons must flow before reaching the outer crust and surface. And the more rock this fluid must pass, the more opportunity is available for gathering radioactively derived helium atoms along the way. The helium concentration in a gas can thus be used as a rough indication of the depth from which this gas has come. Though approximate, this calculation should nevertheless enable us to distinguish between a putative biogenic source at a depth of, say, five kilometers and an abiogenic source at 150 kilometers. A depth-related variant this large should swamp the much weaker variations attributable to differences in the concentrations of radioactive elements in the rocks encountered along the path of escape.

Another line of support for this depth hypothesis is the extent of surface area over which particular gas mixes emerge. The surface area of emergence may reveal the relationship of the depths in which various gases were liberated. A vast geographical spread of an identifiable mix means that those gases must have arisen from a very deep level. Another mix that draws a smaller patch on the surface, but within the area of the first, must have come from a shallower level. When I and a colleague (Marshall Held) analyzed these kinds of data, which had already been recorded for petroleum fields in Texas and Kansas, we discovered that of all the volatile elements or compounds, nitrogen in the form of N_2 and helium frequently would have derived from the deepest levels, natural gas (methane) from the next deepest, and oil with various admixtures of hydrocarbon gases from the next. But in any event, all would come from levels far deeper than the crustal sediments.[9]

Helium enrichment is rarely found in sedimentary rock in the absence of larger amounts of hydrocarbons or nitrogen. Ten percent helium in methane–nitrogen gases is the highest concentration that has been encountered. For helium to be concentrated in oil- or gas-producing regions to much higher concentrations than in the neighboring rocks, nothing other than a mechanical pumping action can be invoked. No chemical action is possible for the inert helium. But why would any pumping action drive helium specifically into petroleum-bearing

areas? The association is so strong that all the world's commercial production of helium comes from oil and gas wells, and the concentration
of helium is often greater by a factor of a hundred in the oil-bearing
areas than in neighboring ground. Even at the shallow depths of farm-
ers' water wells, measurements of helium and methane concentrations
show a close relationship. Measurements of this kind have been carried out in hundreds of data-gathering points in several parts of the
globe (Figure 4.3). The relationship cannot be doubted.[10]

I cannot think of any pumping action that would drive helium from
surrounding regions horizontally just toward a hydrocarbon-rich area.
The only action I could understand that would bring helium into the
hydrocarbon areas would be the diffusely produced helium in the
rocks being swept up over a great interval of depth by a gas that itself
contributed to the hydrocarbon reservoirs.

Because helium derives mainly from ongoing radioactive decay of
uranium and thorium, it is only very diffusely distributed in the rocks.
By itself, radioactively produced helium could not create pressures
sufficient to open pore spaces in the rocks to allow any bulk flow to
occur. Molecular diffusion of helium through the rocks would be the
only mechanism for its ascent. Although molecular diffusion of helium
would be faster than that of any other gas, it would nevertheless be
much slower than bulk transportation. Helium transport therefore
must be driven by another and much more abundant gas that provides
its own motive force both for upward streaming and for generating
pressure-induced pore spaces and fissures along the way. It is the driving force of this other gas that provides the required pumping action
that concentrates helium in hydrocarbon reservoirs near the surface. If
that other gas is a hydrocarbon, it will of course pump the helium it has
picked up into the regions that we identify, at shallower levels, as
hydrocarbon-bearing. This, then, would account for the association of
hydrocarbons with helium.

The test of the hypothesis that hydrocarbon fluids serve as the
upward carrier of helium would be this: If helium could flow without a
carrier fluid, there should be many locations where amounts of helium
had accumulated that were similar to the amounts of helium in some

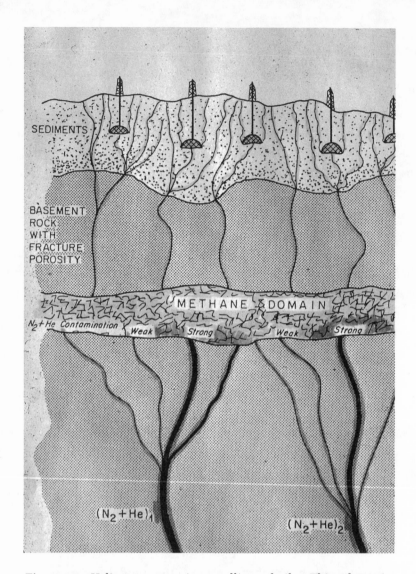

Figure 4.3 Helium transport in upwelling volatiles. This schematic shows how the deep-earth gas theory would account for the helium association with methane. From the deepest levels (perhaps about 300 kilometers), helium produced by radioactive decay is swept into the stream of upwelling nitrogen. At a depth of perhaps 100 kilometers, designated here as the methane domain, nitrogen and helium mix with methane, and all three continue their journey upward. These gases then arrive in the final fields with mixing ratios already determined. The nitrogen–helium ratio is constant over a much larger area, whereas the mixing ratios with methane display individual smaller areas within the first.

gas fields; but in the absence of methane or nitrogen, these accumulations would be pure helium fields. Given the extent of geological exploration, many such fields should have been discovered by now; they would be of great value. Their absence thus supports the carrier-gas concept of helium transport.

In this exploration of evidence confirming the abiogenic theory of petroleum formation, we have examined data pertaining to three important features that can be directly sampled or detected in the outer crust: the phenomenon of reservoir refilling, isotopic and other features of the crust's vast stores of carbonates, and the strong association of helium with hydrocarbons.

The deep earth gas theory is indeed made compelling by just the helium association,[9–16] which has no other explanation. An association of hydrocarbons with specifically *primordial* helium seen in many locations adds further support.[17–27]

But now we have shifted the difficulty to the other side of the argument: How did hydrocarbons attract biological molecules if they came up from depths that are much too great to make or even maintain such molecules? So where did the hydrocarbons originate? If at great depth (and then necessarily at high temperature) we can understand why they gathered up much helium. If they originated in the shallower sedimentary domain we would understand the presence of biological molecules. We seem to have a paradox here: Two types of perfectly secure information, but with explanations for them that conflict with each other. The resolution of this apparent paradox is the subject of the next chapter.

Chapter 5 Resolving the Petroleum Paradox

T he prevailing theory among Western scientists is that petroleum derives from the buried and chemically transformed remains of once-living cells. To contend, as I do, that complex hydrocarbons were primordial constituents of solar system debris out of which the planets formed, and that these hydrocarbons remain in an unoxidized state within the earth's crust and upper mantle even today, is thus a radically contrarious view of petroleum formation. But the problem remains: If complex hydrocarbons found within the earth's crust are not the reworked remains of surface life, why then does petroleum contain the signature of life?

Supporters of the biogenic theory of petroleum formation build their case on four central observations.[1] First, all natural petroleum contains admixtures of groups of molecules that are clearly identified as the breakdown products of complex, but common, organic molecules synthesized by life. These molecules, which appear to be present in reservoirs of petroleum the world over, could not have been built up in a non-biological process.

Second, petroleum frequently exhibits an optical property suggestive of biological activity. When plane-polarized light is passed through a sample of the fluid, the light emerges with its plane of polarization rotated. This rotation implies that molecules that can form with either a right-handed or a left-handed symmetry (in the same sense as we know right-handed and left-handed screws) are not represented by (statistically) equal numbers in petroleum. Rather, one "hand" dominates—which is characteristic of biological liquids, presumably because their molecules have a common ancestry, but this trait is absent in fluids of non-biological origin.

Third, some petroleums are mixtures of hydrocarbon molecules in which those with an odd number of carbon atoms are more abundant than those with an even number. Again, the inverse is also found. It has been suggested that such an odd–even effect can be understood as arising from the breakdown of certain classes of molecules that are common in biological substances that may have contributed to any particular sample. As in the case of the optical effect, no detailed explanation has been put forward. Biology may well be involved, but not necessarily as the source of these molecules.

Fourth, petroleum is found mostly in sedimentary deposits and only rarely in the primary rocks of the crust below. Even among the sediments, petroleum favors strata that are geologically young. In many cases, such sediments appear to be rich in tar-like molecules known as kerogen. These molecules are interpreted by supporters of the biogenic view as having a biological origin, and they are regarded as the source material for any petroleum deposits found in the vicinity. Earth processes are believed to convert these diffusely distributed kerogen molecules into fluid petroleum. Earth processes are also believed to direct the petroleum somehow into concentrated reservoirs within porous sedimentary rock, sometimes concentrated by factors of 100 relative to the distribution of kerogen, and at substantial lateral distances from the kerogen-rich rock considered the source. No theory of this concentration process has been proposed.

The Deep Hot Biosphere Solution

I spent years puzzling over the conflicting evidence of petroleum formation. For reasons explained in the previous two chapters, how could the abiogenic theory be squared with the equally strong evidence of biological activity? As it turned out, the problem had become a paradox only because arguments on both sides contained an unrecognized hidden assumption.

There are no real paradoxes in science; the apparent paradoxes are merely nature's polite way, *sotto voce,* of informing us that our understanding is incomplete or erroneous. With respect to the petroleum paradox, the unrecognized assumption on both sides of the debate was an unquestioned belief that life can exist only *at the surface of the earth.* None of us had considered that a large amount of active microbiology could exist *within the earth's crust,* down to the deepest levels to which we can drill.

That assumption is a vestige of what I've dubbed surface chauvinism, the belief that life is only a surface phenomenon. If we can strip away that assumption, we can entertain the proposition that the biological molecules present in crude oil are not vestiges of surface life long dead, buried, and partially transformed. Rather they are evidence of a thriving community of microbes living out their lives at depth, feasting on hydrocarbons of a deep, abiogenic origin. Once free from the preconceptions, we can open our eyes to the existence of a deep hot biosphere—and one of immense proportion would be needed to account for all the biological molecules in oils around the globe.

For some time before I recognized that the theory of the deep hot biosphere could resolve the petroleum paradox, I had stressed that oils traveling up through the sediments would leach out any biological materials encountered along the way and that such leaching would provide these fluids with biomolecules.[2] It was difficult, however, to reconcile this process with the fact that some petroleum reservoirs have no plausible connection with sedimentary strata in which biological materials might have been buried and thus have been subject to leaching. These problematic oils include those from the very deepest basement rocks in which samples of oils have been found. The solution had to be an abiogenic origin of all petroleum and natural gas, coupled with the extraordinary proposition that a huge microbial biosphere existed at depth, down to at least eight kilometers, (which is the depth at which petroleum in the deepest boreholes has been found). In this view, all petroleum sampled from the ground would have supported an active microbial life because oil is a very desirable substance for various forms of microbiology. We see clearly that where the temperature of the oil is low enough for microbes to flourish, the biological markers are present.

Coupling the abiogenic, or deep-earth gas, theory with the assumption that a deep hot biosphere exists allowed me to interpret the petroleum–helium association as deriving from the long pathways the petroleum fluids had traveled from their deep origins to the outer crust. Having originated far below the depth limit for any biology that could have spiked it with the biomolecules that we observe in it now, petroleum must have traveled up without possessing any of these molecules. Upon reaching shallower levels, where conditions allowed biology to function, the upwelling petroleum quickly became loaded with the great variety of molecular species that a vigorous microbiology could produce at such levels.

Indisputable evidence of living indigenous microbes has been reported in oil wells at depths of more than four kilometers, as we saw in Chapter 2. I believe that all the depths to which our drills can reach, and from which we therefore obtain samples for analysis, are shallower than the transition level below which biology cannot operate. Hence all hydrocarbons will show this type of biological enhancement, but for a very different reason from that assumed by the biogenic theory. Living cells, not just biologically derived molecules, have already been drawn up from deep wells and successfully cultured. The deep hot biosphere is of immerse extent, even though it is limited to pore spaces and fissures within the rock.

Biological Molecules in Non-Biological Petroleum

My theory of the deep hot biosphere thus required that we acknowledge the existence of a previously unrecognized and huge domain of life. It was this assumption that would resolve the petroleum paradox, but it was a very large assumption to make. Corroborating evidence would be required before it could be accepted. That evidence emerged in 1984, in the form of a remarkable paper by Guy Ourisson, Pierre Albrecht, and Michel Rohmer, working at the University of Strasbourg.[3] Although I disagreed with the authors on their main conclusion, that bacteria had *produced* the oil and coal—(With what food? With what sources of carbon and hydrogen?)—the quantities of microbial materials that they reported greatly strengthened my willingness to make the extraordinary assumption of a rich biosphere at substantial depth.

The authors showed that the quantity of biological debris in petroleum was astonishingly large, even though the *proportions* in petroleum were small. A massive bacterial contamination was implied in any case, although this was not the opinion of these authors. The Ourisson team, rather, expressed the conventional view that biology was essential for the production of hydrocarbons. They did not consider that the oils could be food for a prolific microbial life and thereby create the association between petroleum and biology. I responded in a letter published in the same journal in November 1984, writing in part,

> A widespread early bacterial flora may have arisen when hydrocarbon outgassing of the earth provided a source of chemical energy in the surface layers of the crust where oxygen was abundant owing to the photodissociation of water and the loss of the hydrogen to space. Methane-oxidizing bacteria (and possibly also oxidizers of hydrogen, carbon monoxide, and hydrogen sulfide) may have been able to thrive in the crustal rocks. In the course of evolution, photosynthesis, with all its complexity, may well have been preceded as a source of energy by hydrocarbon outgassing. The flora the outgassing sustained gave oil and coal its distinctive biological imprint.[4]

One molecular signature of life in oils came from a group of molecules that the Ourisson team had found and named *hopanoids*. Hopanoids are slightly oxygenated and enriched versions of the hydrocarbon molecules known as hopanes, which contain anywhere from about 27 to 36 atoms of carbon arranged in contiguous rings in a single molecule. The higher-carbon hopanoids contain the extra carbon com-

ponents as a chain added onto the linked rings. Hopanoids are prominent in all of the numerous samples of petroleum that have been tested
for them. This includes samples drawn from sediments of widely ranging ages and from all over the world. And there is no dispute that these
molecules are derived from the membranes of once-living cells.

The amount of hopanoids was huge, the authors argued: "The
global stock of hopanoids alone would be at least 10^{13} or 10^{14} tons, more
than the estimated 10^{12} tons of organic carbon in all living organisms."

Ourisson and colleagues were puzzled, however, by the fact that
whereas living trees and ferns and algae are known to contain hopanoids
at the lower end of the carbon-number spectrum, only bacteria[5] contain
the higher-carbon molecules, such as C_{35} and C_{36}. Another interesting
molecule (a terpenoid) that the Ourisson team found to be common in
hydrocarbons is also present in bacteria known to make their living by
oxidizing methane. The biogenic molecules discovered in natural hydrocarbons throughout the world can all be linked to constituents of bacteria or archaea, and none is linked exclusively to macroflora or fauna.
There is thus no evidence in these observations that anything other than
a substantial microbiological contamination of oils is required to explain
all the molecules observed. And this means, in turn, that there is no evidence that any surface life must be invoked to explain the presence of
these biological molecules in subsurface hydrocarbons.

One need not have waited for hopanoids, however, to cast doubt on
the biogenic theory of petroleum formation. Robert Robinson made the
most persuasive argument more than a decade before petroleum geology had claimed my attention. "It cannot be too strongly emphasized,"
he wrote in 1963, "that petroleum does not present the composition
picture expected from modified biogenic products, and all the arguments from the constituents of ancient oils fit equally as well, or better,
with the conception of a primordial hydrocarbon mixture to which bio-
products have been added."[6]

Quite simply, it is most unlikely that any biological debris could be
degraded into hydrogen-saturated hydrocarbons. Robinson's line of reasoning still stands, and it remains perhaps the easiest-to-understand and
most compelling of all biochemical arguments against the biogenic theory.

Nobody has yet synthesized crude oil or coal in the lab from a beaker of algae or ferns. A simple heuristic will show why such synthesis would be extremely unlikely. To begin with, remember that carbohydrates, proteins, and other biomolecules are hydrated carbon chains. These biomolecules are fundamentally hydrocarbons in which oxygen atoms (and sometimes other elements, such as nitrogen) have been substituted for one or two atoms of hydrogen. Biological molecules are therefore *not* saturated with hydrogen. Biological debris buried in the earth would be quite unlikely to lose oxygen atoms and to acquire hydrogen atoms in their stead. If anything, slow chemical processing in geological settings should lead to further oxygen gain and thus further hydrogen *loss.* And yet a hydrogen "gain" is precisely what we see in crude oils and their hydrocarbon volatiles. The hydrogen-to-carbon ratio is vastly higher in these materials than it is in undegraded biological molecules. How, then, could biological molecules somehow *acquire* hydrogen atoms while, presumably, degrading into petroleum?

Consider too that in oil wells, the average hydrogen-to-carbon ratio increases with depth, corresponding (according to the abiogenic view) to a hydrogen loss through time and during the upward migration of the fluids. Yet a deeper hydrogen reservoir would be regarded as "older" than a shallower reservoir in the same vicinity, given the rule of superposition in geology—that younger sediments are deposited on top of older sediments. Why, then, would the deepest deposits have the highest ratio of hydrogen to carbon?

Equipped with the theory of the deep hot biosphere as the solution to the petroleum paradox, I made an estimate (published in 1992) of the biomass that such a biosphere might support.[7] Let us begin with a presumed upper temperature limit for life of 110°C to 150°C (which at considerable depth would still be well below the boiling point of water). This would place a depth limit for deep biospheric life at somewhere between 5 and 10 kilometers below the surface in most areas of the crust. The total pore space available in the land areas of the earth down to a depth of 5 kilometers can be estimated as 2×10^{22} cubic centimeters (taking 3 percent porosity as an average value). If material of the density of water fills these pore spaces, this would represent a mass of 2×10^{16} tons. What fraction of this might be bacterial mass?

Here the calculation becomes highly speculative. Let us estimate, rather conservatively, that bacterial mass occupies just 1 percent, or 2×10^{14} tons, of the total material occupying the pore spaces. In that case, biomass originating and contained within the deep hot biosphere would be equivalent to a layer of living material that would be approximately 1.5 meters thick if it were spread out over all of the land surface. This would indeed be somewhat more than the existing flora and fauna of the surface biosphere, and it comports with the worldwide estimate of biological debris—hopanoids—calculated by the Ourisson team to be present in all crude oils.

We do not, of course, know at present how to make a realistic estimate of the subterranean mass of material now living. But my rough estimate and that of Ourisson and colleagues indicate that it well could match or exceed all the living mass of the surface biosphere.[8]

The Upwelling Theory of Coal Formation

The dominant view in Western countries is that crude oil and natural gas derive from biological debris reworked by geological processes, as we have seen. In contrast, the abiogenic view, coupled with the theory of the deep hot biosphere, is that liquid petroleum and its volatiles are not biology that has been reworked by geology but *geology that has been reworked by biology.*

The evidence we now have of biological activity in petroleum is the cellular remains of microbes that fed on hydrocarbons, some at a depth of perhaps ten kilometers, where it was previously thought that no biology could be present. This microbial activity is not just something that happened in the remote past; it is still going on. Oil and gas reservoirs are still filling and still venting to the surface, and the denizens of the deep hot biosphere are still degrading fresh supplies of oil into carbon dioxide and other excretory products as they live, reproduce, and die.

The abiogenic theory well accounts for many spatial and chemical features of these hydrocarbon reserves that the biogenic theory has not been able to explain. And the biological molecules detected in crude oils are explained by the deep hot biosphere. Crude oil and natural gas are thus by no means "fossil fuels," as they are often termed. But surely I must make an exception for coal, one would think.

No. I contend that although peat and lignite do originate from undecomposed biological debris, black coals do not. In my view, black coals form from the same upwelling of deep hydrocarbons that accumulate as crude oil and natural gas. With coal, however, the hydrogen component has been further driven off, leaving behind a greatly carbon-enriched, hydrogen-impoverished hydrocarbon. How could coal form in this way? What empirical evidence is there for this contention as opposed to the biogenic theory?

Many people think that the origin of coal is completely understood. This is not the case. What has happened is what happens only too often in science: An unsatisfactory explanation is accepted because no more satisfactory explanation turns up over a long period of time. The biogenic theory of coal formation demands the assumption— unwarranted in my opinion—that lands all around the globe formerly supported vast stretches of swamp forests in which generation upon generation of tree ferns (during the Paleozoic) and conifers (during the Mesozoic) fell into oxygen-depleted waters, thus preventing decomposition. Moreover, these "coal swamps" occupied down-warping regions, in which thousands of feet of overburden, sometimes alternating with swamp conditions, pressed down upon the buried plants as the eons passed. Pressures and temperatures prevailing at depth, given sufficient time, would then somehow transform biological molecules into black coal.

Early investigators of bituminous coal (beginning in England, for example, as far back as the 1850s) found a lot they could not explain about the substance's composition. Because there were some fossils in the coal, and because life on the earth is carbon-based, a biogenic theory seemed quite plausible and seemed the best course to pursue in the absence of an alternative. Nevertheless, the biogenic theory was at a loss to account satisfactorily for most or all of the situations in which coal is found.

It is indeed true that coal sometimes—though by no means always—contains some fossils, but those fossils themselves create a problem for the biogenic theory. First, why did the odd fossil retain its structure with perfection, sometimes down to the cellular level, when other, presumably much larger quantities of such debris adjoining it were so completely demolished that no structure can be identified at all? Would it not be strange for one leaf or twig to have its shape perfectly preserved and for all other leaves and twigs in the same assemblage to be turned (by high pressure) into a uniform mass of almost pure carbon? Second, fossils are sometimes filled almost solid with carbon without being deformed. Every cell of the plant seems to have been filled with the same coaly material that forms the bulk of the coal outside the fossil. How did the coaly material enter the structure of the fossil without destroying it? Such coal fossils seem to be filled with carbon in the same way that petrified wood is filled with silica.

Silica-rich petrified wood is universally believed to have been introduced by the flowing through of aqueous fluids rich in dissolved silicon dioxide. Over time, silicon dioxide—quartz—is deposited, crystallizing in ways that conserve the cellular structures without conserving any of the cellular contents. Why should not the same sort of process, involving a very different fluid, have been at work in the formation of coal and its fossil inclusions? The "coal" in the cells of the plant must have entered as a fluid, and presumably it was the same fluid that laid down the surrounding matrix of coal.

If not only crude oil and natural gas are a gift of the deep crust or the mantle, but coal is too, how might coal actually form?

To begin with, simple chemistry and physics tell us that hydrocarbons will suffer a loss of hydrogen on their way up through the crust. Why is this so? First, any opportunity for a stray (or microbially catalyzed) oxygen atom to interact with a hydrocarbon fluid of any sort will result in that fluid losing two hydrogen atoms for every oxygen atom encountered, thus generating water. This represents nothing more than a drive toward chemical equilibrium. The carbon-to-hydrogen ratio of the fluid will then rise, with other appropriately charged atoms (such as nitrogen and sulfur) taking the place of hydrogen in the molecular structure or, more commonly, with double bonds replacing single bonds in the carbon chains or rings that must accommodate hydrogen loss.

As we know all too well from the faults in the engines of our cars, oxygen atoms preferentially strip off almost every hydrogen atom in a hydrocarbon fluid before going to work on the carbon atoms, which are more difficult to oxidize. The result: less than fully oxidized carbon (carbon monoxide, CO). A sign of even less efficiency is the unoxidized black stuff ringing the head of a spark plug drawn from the engine of a poorly tuned car or pouring out of the exhaust pipe of a heavily laden truck shifting gears. This is nearly pure carbon, or soot.

Generally, the "heavier" (more carbon-rich) the fuel, the greater the chance of incomplete combustion. In order to convert all the carbon as well as the hydrogen in a hydrocarbon fluid into molecules of the highest oxidation state, sufficient oxygen and sufficiently high temperatures must be provided. A paraffin candle flame, for example, is hot enough to oxidize the hydrogen, but it is not hot enough to oxidize all the carbon. Soot is, in fact, the intentional product of the carbon-black industry, which incompletely burns natural gas in a low-oxygen, low-temperature environment in order to produce soot that can be sold as printer's ink.

Oxidation is not the only cause of hydrogen loss on the way up from the earth's depths. Complex hydrocarbons forged at depth would be unstable at near-surface pressures even if they would be stable at the pressures that prevail at their point of origin in the upper mantle, perhaps 200 or 300 kilometers beneath the earth's surface. In the upper rocks, and away from volcanic influences, temperatures are too cold to break the molecules apart forcibly, but there will nevertheless be a gradual dissociation of hydrogen from the carbon, as the hydrocarbon mix adjusts gradually to the lower pressures of shallow depths.

The existence of diamonds—crystals of pure carbon—gives us several very significant items of information about the circumstances at depths of more than 100 kilometers. (This important topic will be examined in more detail in Chapter 7.) The pressure necessary for carbon to adopt this crystallographic form has been reliably determined to be 35 to 40 kilobar (1 kilobar is equal to 1000 times the atmospheric pressure). Because the pressure at any depth cannot exceed the value given by the overburden weight of rock, diamond formation could not occur at depths less than 100 to 150 kilometers. By studying diamonds, we can thus learn something of the conditions at such great depths.

First, a process has to be at work that will concentrate high-purity carbon. Only a flow of a liquid that carries carbon can do this. Pore spaces must exist down there, and fluids must flow through them that can shed pure carbon. Second, small impurities that exist in some diamonds, in the form of inclusions of fluids at a pressure similar to that needed for the formation of diamonds, can be considered samples of fluids that occur at such depths. Among those are methane, other light hydrocarbons, and CO_2. This answers the question of the depth at which some unoxidized carbon compounds are stable in the earth: It is at least 100 kilometers but may be much more. I presume that the dissociation of some hydrocarbons is the origin of the clean carbon of the diamonds. Even diamonds made in a pressure–temperature domain where they are stable become unstable at the low surface pressures. Diamonds are not forever, but they are for long enough, only because they are supercooled and do not have the energy to change their crystal configuration to the low-pressure form, which is graphite. Similarly, most hydrocarbon molecules will enter an unstable domain as they rise toward the surface.

To get a sense of how hydrocarbons spontaneously reconfigure their molecular mix under a substantial change of pressure and temperature, I directed a graduate student to study chemical changes in a sample of propane (C_3H_8) subjected to a simulated depth environment of 475°C temperature and 4000 atmospheres pressure, which corresponds to a depth of about 10 kilometers in much of the earth's crust. After only six hours, the sample had rearranged itself into a mix ranging from C_1 to C_5, maintaining the input ratio of carbon to hydrogen (Figure 5.1). This shows that hydrocarbon molecules can be assembled without the intervention of life; it also shows that the ratio of the vari-

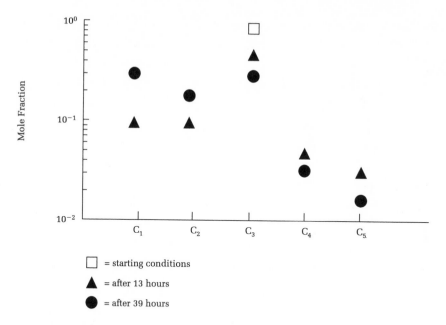

Figure 5.1   Spontaneous modification of hydrocarbons under a
change of heat and pressure. In this experiment a nearly pure quantity of propane (C_3), sealed in a gold-lined vessel, was subjected for 6
hours to a temperature of 475°C and a pressure of 4000 atmospheres
(the latter simulating a depth of about 10 kilometers). Under those
conditions, some of the propane spontaneously reconfigured into
both lighter and heavier hydrocarbons, thereby maintaining the
hydrogen-to-carbon ratio in this closed system. The final mix ranged
from methane (C_1) to pentane (C_5). SOURCE: Thomas S. Zemanian,
William B. Streett, and John A. Zollweg, 1987, "Thermodynamic calculations, experiments, and the origin of petroleum" (unpublished
results). A similar but more complicated experiment is reported in
Thomas Gold *et al.,* 1986, "Experimental study of the reaction of
methane with petroleum hydrocarbons in geological conditions,"
Geochimica et Cosmochimica Acta 50: 2411–18.

ous light hydrocarbon molecules seen in oil and gas wells is determined by the pressure–temperature relation along the path of the
ascent.

Sequential loss of hydrogen is the primary reason why so many
petroleum fields are configured in a layer-cake fashion: Vast methane

deposits at the greatest depth, light oils higher up, and the heaviest oils on top (though each pocket may be capped with some amount of methane). In some fields, the most carbon-rich and topmost hydrocarbon is not crude oil; crude oil is not always the end of the sequence. Rather, above the oil layers may be black coal. The blacker the coal (from bituminous to anthracite), the greater the loss of hydrogen and the greater the resulting carbon-to-hydrogen ratio.

What about the biological molecules detected in coal? The presence in coal of the same sorts of hopanoids—molecules attributable to bacteria—that are found in crude oil is strong evidence that the same microbes are dining or have dined on hydrocarbons in coal layers as in oil reservoirs (Figure 5.2). Under the biogenic theory, however, the strong similarity in hopanoid species in coals and oils is difficult to explain.[9] This is because proponents of the biogenic theory regard coal as the altered remains of land plants and oil as the altered remains of marine biological debris, yet it would seem improbable for nearly the same microbiological material to be found in both. If coal was an end product of oil, then this coincidence would be explained.

Wherever microbiology has played a catalyzing role in the conversion of a hydrogen-rich hydrocarbon into a hydrogen-poor hydrocarbon, the product can, to some extent, be considered a biological creation—but one produced by an underground ecology feeding on abiogenic petroleum fluids. In the genesis of coal, however, it is unclear whether and to what extent microbes of the deep biosphere are involved. The process of hydrogen loss may or may not be assisted by microbial action. Either way, positive feedback can be at work once the first atoms of pure carbon are generated. A deposit of solid carbon acts as a catalyst for further deposition of carbon from methane or other hydrocarbons. Where other circumstances, such as temperature and pressure, would come near to causing dissociation and subsequent deposition of carbon, the presence of some carbon will initiate the process. This means that in an area of upstreaming hydrocarbons, there will be a tendency for carbon deposits to grow to great concentrations, because their very presence is instrumental in laying down more of the same stuff.

I came to know this well from an experiment carried out in my laboratory. We began with a transparent tube of fused quartz, partially sur-

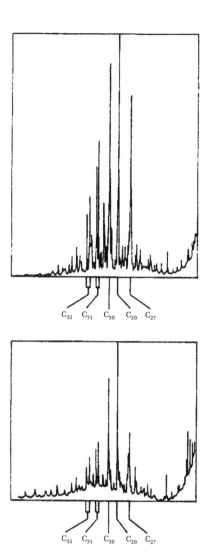

C_{32} C_{31} C_{30} C_{29} C_{27}

C_{32} C_{31} C_{30} C_{29} C_{27}

Figure 5.2 Similarity in hopanoid (biological) molecules detected in a coal sample and an oil sample. The upper chromatogram was obtained from a Lorraine (France) coal residing in strata dated at about 300 million years. The lower chromatogram comes from a heavy crude oil that is found in strata of the Aquitaine Basin, also in France, dated at about 150 million years. Comparing these chromatograms shows that coal and oil had a similar complement of bacteria, depositing the unusual form of the biological debris. The judgments of the ages are those for the containing rock, whereas the carbon may have been laid down later. SOURCE: Guy Ourisson, Pierre Albrecht, and Michel Rohmer, 1984, "The microbial origin of fossil fuels," *Scientific American* 251(2): 44–51.

rounded by an intense electric heater, but leaving space to see the interior. Methane was blown through the tube as the temperature rose. At around 800°C, a black speck suddenly appeared on the inside wall, and within a fraction of a second a black streak developed, starting from the initial point and widening as a triangle in the downstream direction. Significantly, carbon did not appear in a scattered pattern inside the tube. Rather, once the first speck had been generated, all subsequent deposition created a single, expanding mass; carbon was deposited very quickly after the first grain appeared.

In conclusion, I believe that coal may be formed by both abiotic and biotic processes. What distinguishes this theory from traditional theory is that coal is postulated to be derived from a source upwelling from the depths rather than a deposit sinking from the surface. We might therefore refer to this as the *upwelling theory*. The carbon has entered from below as a carbon-bearing fluid such as methane, butane or propane, and thus it could penetrate into the cells of any plant fossils that were present in the flow path. After that, a continual loss of hydrogen would gradually bring it closer to the consistency we call coal. Black, hard coal is a product of entirely subsurface processes; it has nothing to do with the surface biosphere. It has nothing at all to do with photosynthesis. Such coal is not the stored energy of the sun.

Evidence for the Upwelling Theory

Evidence in favor of this upwelling theory of coal formation is various and, in my view, compelling. Perhaps the strongest refutation of the traditional theory of coal formation can be found in the paucity of mineral ash in most black coals. Some coal seams are more than 10 meters thick, yet the mineral content may be as low as 4 percent. The bulk of the material is just carbon, with a little hydrogen, oxygen, and sulfur mixed in various compounds. For a swamp to lay down enough carbon to produce such a seam, it would need to have grown to a depth of more than 300 meters, with a mineral content in that volume of less than 1 percent. No such swamps exist today, and even if they once existed, it seems unlikely that plants would grow in such circumstances.

The ratio of minerals to carbon in any present-day accumulation of plant debris is a very much higher one, and accumulations of the quantities of biomass carbon necessary to account for major coal seams are not found anywhere. There is no reason to invoke an environmental process (terrestrial coal formation) and a breadth of suitable surface environment (vast forested swamps) that have no analogs in today's world. A more parsimonious theory surely is in order, especially inasmuch as we understand that a large quantity of carbon must have been delivered to the surface over geological time.

The upwelling theory, in contrast, can account for the low mineral content of coal, and it avoids the need to posit a type and scale of environment that once occurred but exists no longer. I believe that substantial coal formation is not just a thing of the past. It is happening today. We do not recognize it simply because coal is forming largely as incremental additions to existing coal deposits. Not only the oil and gas fields are recharging but the coal deposits also, only at a rate too slow for us to recognize.

The upwelling theory well suits the dimensions of massive coal deposits and explains the small amount of mineral ash contained therein. Perhaps perfectly "ordinary" biological deposits in a sedimentary layer of a helpful porosity, with normal admixtures of minerals, would act as a starter. Some of the upstreaming hydrocarbons would be dissociated there; the fossils in that rock would be filled with carbon, and as more carbon accumulated, it would stimulate further accumulations of carbon. In the end, the quantities of carbon attributable to the original plant material may be an insignificant fraction, and the carbon-to-mineral ratio may have reached values that never occur in surface vegetation. The commonly occurring vertical stacking of coal seams then merely attests to the area being one in which hydrocarbon fluids have been outgassing over extended periods and in which the circumstances have been mildly in favor of dissociation and carbon deposition. This also accounts for another observation much researched in recent times: Coal often seems to produce large and commercially valuable quantities of methane gas, especially when the ground-water pressure surrounding a coal seam is lowered. It is then said that this methane must have been resident in the coal. But methane moves rather freely through coal, and no reason can be found why it would have become concentrated and remained there for periods as long as the age of any particular coal seam. It is more likely that the presence of the coal is a good indicator for the upwelling methane.

Whereas the ash content in coal is far less than the biogenic theory would predict, concentrations of trace minerals may be far greater than the traditional view can explain. By far the greatest human-made contributor to radioactive pollution is not leakage from the wastes and cooling water of nuclear power plants but uranium-rich plumes from the smokestacks of coal-fired power stations. In addition to uranium, metals such as mercury, gallium, and germanium are often found concentrated in coal, much beyond normal sedimentary levels. These are not metals one might expect to have been concentrated by plants. Therefore, advocates of the biogenic theory of coal formation presume that some coal seams must have acted with the same agency as the charcoal in cigarette and in water filters: extracting metals passing through. But in most cases of extreme concentrations, it is difficult to see how ground water could have transported enough of the substance through the coal, even if all that passed had been arrested there. In contrast, the upwelling theory holds that coal, like petroleum, is formed from hydrocarbons brought up from the deeps, leaching minerals throughout their journey.

Another anomaly that is difficult for geologists to explain through the biogenic theory is the presence of coal seams in places where they ought not to be and at inclinations they ought not to take. Most commercially mined coal seams are layered between sedimentary strata, but many coal deposits in the world are not. Coal that is interbedded with volcanic lava and without any sediments is known in several volcanic areas, most notably in southwestern Greenland.[10] There coal is found in close proximity to large, lava-encrusted lumps of metallic iron, not far from mud volcanoes burbling methane and from a rock face that frequently has flames issuing from its cracks.[11] Another notably non-sedimentary deposit is located in New Brunswick, Canada. There a coal

called Albertite fills an almost vertical crack that goes through many horizontally bedded sedimentary layers. It was mined in the last century,
but the difficulty of mining a nearly vertical seam caused the operation
to be curtailed.[12] The biogenic theory can offer no remotely plausible
causal explanations for these and other anomalous coal environments.

Then, too, it is not uncommon to find lumps of carbonate rock
within a coal seam and, upon breaking them open, to find fossils containing wood—not black but light in color—and showing no signs of
turning into coal. Similarly, it is reported that in the coal of the Donetz
Basin of the Ukraine can be found fossilized tree trunks that span
through a coal seam from the carbonate rock below to that above. Those
fossils are coalified where they are within the coal seam and are not
coalified where they are in the carbonate.[13]

Many investigators have remarked on the numerous inconsistencies that one sees if one wishes to interpret the coal as a result of
swamp deposition in the locations in which the coal is now found.
H. R. Wanlass, for example, was puzzled by the presence in some coal-
bearing regions of interbedded clay layers only one or a few inches
thick that extended horizontally through the coals, unbroken over distances of several hundred miles. He therefore judged there to be "sufficient objections to all proposed theories of the origin of these clays to
make each seem ludicrous."[14]

The geographical distribution of coal deposits poses another problem for the conventional theory. It is assumed that oil and coal are the
result of completely different types of biological deposits laid down in
quite different circumstances and, in many regions where both occur, at
quite different times. Biological debris from marine algae is usually
invoked for the formation of crude oil, and terrestrial vegetation for
coal. No close relationship between the geographical distributions of
the two substances would thus be expected. But in fact, as the oil and
coal maps of the world have been drawn in ever-increasing detail, a
close relationship has become unmistakable. The coal and oil maps of
southeastern Brazil are striking in this respect (Figure 5.3). Indonesia
presents another such example; local lore among those who drilled
there for oil was "Once we hit coal, we knew we were going to hit oil."

OIL & GAS   COAL

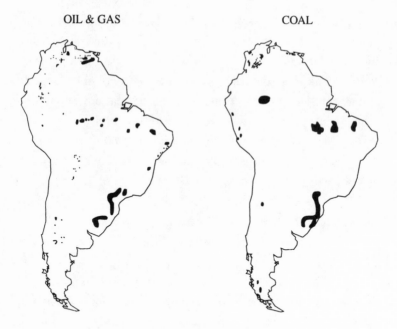

Figure 5.3   Overlap in the distribution of coal and oil in eastern Brazil. Many other such areas of overlap exist, which presents problems for the biogenic theories of coal and oil formation but is readily explained by the abiogenic theory. SOURCE: oil map adapted from International Petroleum Encyclopedia, 1994, p. 85; coal map adapted from a commercial atlas by H. M. Goushu Company, San Jose, Calif.

Coal on top and oil below is such a common feature that chance cannot possibly account for it. In Wyoming, some coal is actually found *within* the oil reservoirs. In many sedimentary basins, including the San Juan Basin of New Mexico and the Anadarko Basin of Oklahoma, coal directly overlies oil and gas (Figure 5.4). Alaska, Iran, Saudi Arabia, the Ural Mountains—all known for their oil fields—also possess large amounts of coal. The same is true of many other major oil-producing areas, such as Venezuela, Colombia, and the Pennsylvania section of the Appalachian Mountains.

Consider too that some coal fields contain and yield more methane than could possibly be produced by the existing coal. Coal that has not

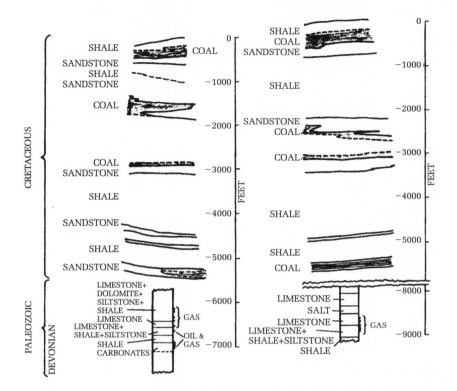

Figure 5.4 An example of the vertical stacking of gas, oil, and coal
deposits in San Juan County, New Mexico. The diagram at the left is
a cross section of the Hogback Field; at the right is the Barker Creek
Field.

yet "blackened" into nearly pure carbon would be expected, under the
biogenic theory, to relinquish those hydrogen atoms slowly, probably
in the form of methane. But if methane is being generated by the coal
itself, rather than upwelling from an even greater depth, it should be
present in very limited supplies. This is not always the case—to the
consternation of coal owners and miners. Even with a very fast,
enforced air flow, many coal mines are plagued with methane explosions. Coal mining on Hokkaido, Japan's north island, has come to a
standstill because even these, the world's best-ventilated coal mines,

could not avoid major explosions. The explanation for the surfeit of methane, to my mind, is that methane from the very source that created the coal deposit is still streaming up. Coal is still forming!

An Exemption for Peat

"What about peat, and what about lignite?" I hear my critics retort. "Surely you do not claim that these are abiogenic!"

No, I do not so claim. Rather, peat and lignite (the latter being a "brown coal" in which the structure of the original plants can still be seen) represent a most interesting partnership between biogenic and abiogenic carbon sources.

There is a subtle connection between peat and lignite on the one hand and oil and coal on the other. Peat and lignite give clear evidence of having been formed by plants in locations where the usual processes of decomposition were prevented from functioning and, where, therefore, the carbon and other components of the plants were not returned to the atmosphere. One way in which this can happen is commonly discussed. If enough plants sink into a pool of water and thus transform it into a stagnant marsh or swamp, then the rate of absorption of atmospheric oxygen will be low. Once even a small amount of the plant debris is decomposed by anaerobic microbes, chemical conditions in the pool may become so hostile that any further decomposition is prevented. The carbon content of the plant fibers will not be turned into carbon dioxide, which would escape, but instead will leave behind a carbon sludge or a carbonaceous and fibrous sponge of materials that will survive for a long time, while also holding up the flow of water.

The anoxic situation in the swamp may often be due to the rapid growth of bacteria plundering any available oxygen atoms in order to burn, for their metabolic needs, abiogenic methane upwelling from below. Because methane is such a desirable food, methanotrophic microbes will outcompete those that would otherwise use oxygen to attack the plant debris, the cellulose and lignin molecules of which many may be particularly resistant to attack. A swamp will then be created from all the plant material that has accumulated and not yet decomposed.

Peat-forming conditions may also arise under quite different circumstances from the swamp example. A peat bog need not be nestled in a bowl that has no natural outflow of water. A peat bog may also be capable of holding water in thick layers of plant debris for a long time without any assistance from topography. Locations have been found in Switzerland where a patch of peaty terrain—that is, a patch of very soft ground vegetated by the same flora that is characteristic of peat bogs—occurs on steep hillsides, along fault lines that run transverse to the slope of the hill. There would seem to have been no impediment to the flow of water, but a measurable output of methane can be detected along the faults. In my view, methane outgassing is therefore likely to create peat and lignite deposits in regions overlying a strong flow of hydrocarbons. I presume this is indeed the explanation for the not uncommon presence of peat and lignite fields on the surface overlying productive oil and gas fields.

Another observation corroborates this presumed causal association between peat and a significant source of methane outgassing. I had measurements done of the gases emanating from a large commercial peat field in Canada. The results were astonishing. In this peat field, the gases just below the surface were greatly enriched with methane, as is commonly the case in such environments—a condition that supporters of the biogenic theory of course attribute to the presence of methane-excreting bacteria feeding on the plant debris in an oxygen-poor environment. But the gases were also enriched with all the other hydrocarbon gases from C_2H_6 to C_5H_{12}. This mix is not normally produced by plants in any of their stages of decay. Quite simply, microbes do not excrete pentane while decomposing carbohydrates.

Curious, and with my upwelling theory in mind, I requested that a hole be drilled well outside this Canadian peat field into the local soil, which contained no peat-like material. The gases drawn from this location proved to be very similar to the gas composition in the peat field itself. The whole area showed the same signs of hydrocarbons. This indicated to me that most of the gases detected within the peat field had in fact entered from below and thus were similar to the gases along the same fault line. Why in this region of outgassing were some patches

peaty whereas others were not? I suppose that differences in the
cleansing effects of various rates of ground-water flow may have been a
factor.

Personal experience with one's own senses—supplementing experience somewhat removed by way of technology or altogether removed
in the library—provides a powerful stimulus for questioning received
views. I have a vivid memory of one such experience in Switzerland. I
was walking along a fairly steep hillside just above a small brook. The
ground was covered not by vegetation but by a slimy mud. A colleague
who was guiding me bent down and stuck his five fingers quite arbitrarily into the mud. He then took out a cigarette lighter and swung the
flame around the holes that his fingers had made. It all lit up, resembling the gas burners on a cooking top in the kitchen! Many others have
apparently had this experience as well, though in another country. I
recall hearing that in a clay field near Oxford, England, the workmen
mining the clay made use of a similar situation to cook their lunches.

Peat and lignite are clearly biological materials, but the reason for
their accumulation may well lie in the circumstances created by non-
biological hydrocarbons that happen to upwell from below and that
may also add more carbon than contained in the plants involved. There
are many locations where one may suspect such a conspiracy between
surface biology and the deep earth. Large peat deposits of Sumatra
reside above oil- and gas-rich regions. Some lignite deposits (for example, those on the north shore of the Straits of Magellan on the Atlantic
side) have commercial deposits of oil and gas just underneath them.
The neighboring Tierra del Fuego—Land of Fire—may have been so
named by Magellan when he saw flames issuing from the ground. This
phenomenon has been incorporated into folklore with frightening tales
woven around the ever-so-real presence of "swamp gas" that may
ignite spontaneously.

Crucially, the black coals do not grade smoothly into the brown
coal of lignite and thence into peat. Rather, there is a sharp discontinu-
ity between the black and the brown—and, to my mind, a sharp dis-
continuity between the circumstances of their genesis as well. Black
coals are the progeny of the deep earth, shaped and glazed by a deep

biosphere feeding on a flowing stream of edibles. In contrast, lignite and peat are the offspring of the surface biosphere—solar energy that has been captured and put into temporary storage, but often held stably, thanks to a bath of hydrocarbon gases flowing up from below.

What about kerogen, though? Kerogen is a tar or coal-like material found as small specks diffusely distributed in various rock strata. Like petroleum, it has never been brewed in a beaker, starting with biological components of any variety and subjected to temperatures and pressures of whatever degree. Whenever kerogen is found near a deposit of petroleum, it is declared to be the "biological source material" that has given off the petroleum found nearby. If kerogen is not in the vicinity, then the petroleum is presumed to have migrated, perhaps a great lateral distance, from a "source" rock that at one time surely did contain kerogen. This explanation of petroleum's origin is, in fact, central to the biogenic theory. But how could a concentrated petroleum reservoir coalesce from a quantity of hydrocarbons that previously had been distributed sparsely in a much larger volume of rock? No explanation for this curiosity has been offered. Because there are indeed chemical and isotopic similarities between kerogen and the neighboring petroleum, adherents to the biogenic theory claim this fact in their favor. But why should the kerogen and the oil in the region not have formed from the same upwelling stream of hydrocarbons?

Thus many scientists who attempt to understand coal seem to have fallen into the rut of the nearest convenient theory. They explore the terrain of this rut very effectively, down to the minutest feature within the walls, yet they will not climb out for another look. "You cannot argue with a fossil," was a remark thrown at me during a lecture I gave on this subject. It is true that you cannot dispute the biological nature of the fossil, but certainly you can think anew about what its presence implies for the material surrounding it.

Chapter 6 The Siljan Experiment

B y the early 1980s I was convinced that the abiogenic theory of petroleum formation was substantially correct. I also knew that in order to solve the petroleum paradox, the abiogenic theory must be supplemented with a theory of what I called the deep hot biosphere. I knew, however, that such views pertaining to the widespread existence of methane and other hydrocarbons deep within the earth's crust would not be taken seriously in the West unless I could offer a clear, practical demonstration of its validity. The theoretical arguments and indirect empirical evidence presented in the previous three chapters would not, in themselves, be sufficient to overturn the reigning paradigm. Rather, I would need to prove that hydrocarbons do indeed exist at a depth and in a type of rock for which the biogenic theory could offer no explanation. I would thus need to generate interest in drilling for oil at a site that would be regarded, under the prevailing view, as among the worst of all possible prospects.

Discovery of oil or gas in even small quantities at such a location might be persuasive, but commercial production in such places would be better. Probably nothing short of this scale would turn heads. Success in this regard would be of more than scientific value. It would be a feat of enormous economic importance—first because the prognosis for future energy supplies would have great influence on oil and gas economics, and second because new thinking and new exploration techniques would be indicated, which could then be used to make new discoveries all around the globe. I would, however, have to convince parties with money and expertise that there was a strong possibility that the areas now known to be rich in oil and gas are not the only ones. I would have to make a sound case that this valuable commodity would prove to be far more widespread around the globe than had been previously supposed and that the dependence of many countries on oil importation would, as a consequence, be greatly reduced.

How could I go about finding hydrocarbons, possibly in commercial quantities, in new places—places that would be quite unexpected according to the conventional theory? How could an academic like myself suddenly become an entrepreneur on the scale necessary to prospect and drill in a place of my choosing—or at least guide a relatively technical operation of that kind?

Sweden seemed to me to be a particularly favorable location for such an experiment. I had traveled extensively in that country years earlier, and I had observed that in many places where granitic bedrock was exposed, cracks in the rock were filled with a substance that looked like tar. When I asked Swedish geologists to explain why tar would be coming up through the granite, they told me the following story. There must once have been a thick layer of sediment overlying most of the bedrock of Sweden, and the organic materials in that sediment produced oil. Cracks that developed in the bedrock underneath sucked this oil downward, and now, millions of years later, it is seeping up again.

This explanation made no sense to me. Water in biological debris in the sediments would surely be far more abundant than oils, and we have all observed that oil floats on water. How, then, could oils be the chief component to penetrate downward into the cracks of the bedrock? This consideration had weighed on my mind for a number of years and, in itself, had seemed to make a good case for the abiogenic theory. In my view, Swedish bedrock represented an incomplete barrier for hydrocarbons *upwelling from below* to reach the surface.

Sweden would also offer an advantage as a test site, being a prosperous and technologically advanced country that however imported nearly all the fuels it required. Drilling into the bedrock of Sweden and finding commercial quantities of the oils responsible for the tar seeps at the surface could point to an ideal energy solution for that nation's inhabitants. In turn, for me, drilling into the igneous rock meant that sedimentary materials could be avoided completely, and so no biological origin could be stipulated if oils were in fact found at depth. The abiogenic theory of petroleum formation would thus be confirmed in the field.

Drilling in Swedish Granite

An opportunity arose unexpectedly in 1983 when I received an invitation to spend a day in Stockholm explaining my ideas to senior officials of the Swedish State Power Board (Vattenfall), an invitation that had been mediated by a lawyer friend in Washington who knew people with Vattenfall. As I now read through the text of my presentation, delivered fifteen years ago, I see that my predrilling arguments were formulated then much as I would formulate them now. Here are some passages from this presentation that were reprinted, in Swedish, in a daily newspaper.[1]

"Natural gas (mainly methane), as well as all natural petroleum, was thought to be invariably of biological origin on the earth. On that basis, the ground of Sweden, composed almost entirely of primary rock and not of sediments, could not come under serious consideration as a source-material for hydrocarbons. The numerous seepages of methane, tars, and oils that occur in the bedrock of Sweden have long been known as a geological puzzle in that context, and various attempts at explanations have been put forward. We now know that the deep borehole on the Kola Peninsula in the far north of European Russia [drilled by the Soviets near the city of Nikel] in similar crystalline rock finds methane at a depth of eleven kilometers as one of the principal gases in the cracks. No explanation in terms of a downward seepage of surface biological material would seem adequate there.

"Sweden would then be seen as a mass of crystalline rock [rock crystallized from a volcanic melt] obstructing the upward flow of gases and liquids from the deep layers below, which in this area of the globe appear to be particularly rich in hydrocarbons. The surrounds of Sweden all show high hydrocarbon levels. The Norwegian trench, stretching from the Dutch coast up to the North Cape and beyond, constitutes the escape route on one side. The Kola Peninsula, as well as the Baltic states, all have evidence or even production of methane. In Sweden, the only escape routes are fractures in the basement rock.

"In most cases crystalline rock does not develop sufficient porosity to hold deposits of oil or gas in commercial quantities. One therefore has to search for locations where the rock is thoroughly smashed and converted to porous rubble. There are several areas in Sweden that can come under consideration, but by far the most prominent is the Siljan Ring [located near the city of Rättvik in central Sweden; Figure 6.1]. Three hundred sixty million years ago a large meteorite struck there and produced a crater some 44 kilometers in diameter. An impact of this magnitude would shatter the crystalline rock all the way through the crust of the earth, and when the ground readjusts its level after the event, the entire interior of the crater will be a region of porous rubble, down to a great depth. This fortunate event created not only the beautiful lakes of the Siljan Ring, but it also left a deeply fractured and porous region in the interior of the circle, in which fluids from below could ascend and collect [Figure 6.2]. The Swedish Geological Survey has already established that the rock is indeed shattered in the region, and a gravity survey is consistent with the expected level of porosity. The interior of the Siljan Ring thus represents a possible reservoir of truly enormous size by any standards.

"Shallow levels in porous rock do not contain exploitable concentrations of gas unless there is a particularly tight caprock to hold it down. This may exist, but we have no reason to expect it in Siljan. At deeper levels the situation is different. The rock compresses under the weight of its overburden and tends to hold down regions in which a high-pressure gas has held open the pores. This type of impediment to the upward flow of gases generally arises at a depth of between three and five kilometers in sediments but must be expected to occur at deeper levels in the harder granitic rock. In the Soviet borehole in Kola, a sudden change to an expanded fracture porosity was seen at a depth

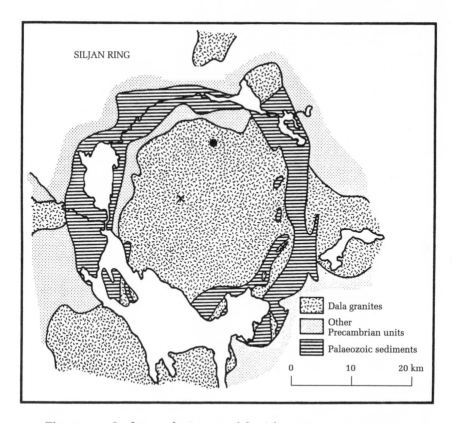

Figure 6.1   Surface geologic map of the Siljan Ring impact structure. The white areas are lakes. The sediments are nowhere deeper than 300 meters. Two deep wells were drilled: The first at Gravberg, to a depth of 6.7 kilometers (indicated by black dot), the second at Stenberg, to a depth of 6.5 kilometers (indicated by a cross). SOURCE: Geological Survey of Sweden.

of seven kilometers, and this probably represents the depth at which this type of rock will crush. Below this level there will then be porosity domains held open by the gas pressure, if indeed gases have been coming up from regions of yet much higher pressure far below.

"One may be lucky and find caprocks holding oil or gas at shallower levels. If not, one has to drill to a depth of seven or eight kilometers, still well within the range of modern techniques. If it has been established that we are indeed dealing with an area of methane outflow from deep levels, then there is every expectation of finding the occluded layers filled with the gas, the outflow at the top representing

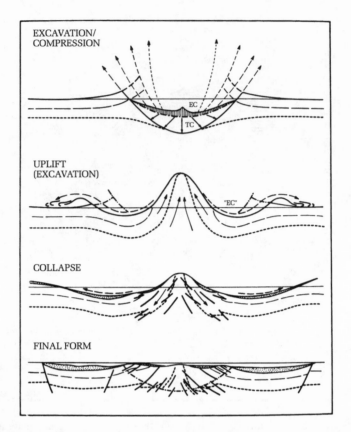

Figure 6.2 Formation of an impact crater such as the Siljan Ring. Note the numerous fractures that would facilitate upward migration of deep fluids. *After* R. A. F. Grieve, 1998, "The formation of large impact structures and constraints on the nature of Siljan," in A. Boden and K. G. Eriksson, eds., *Deep Drilling in Crystalline Bedrock,* vol. 1 (New York: Springer-Verlag), 330.

in fact the spill-over. The size of the area would make it a truly gigantic gas field by any standard.

"A deep hole to eight kilometers may cost on the order of $25 million (U.S.). In many successful areas it has been necessary to drill several holes before establishing good production. The oil and gas business demands a high entrance fee, but it can pay rich rewards. Perhaps a sum of $25 million dollars has to be risked in the first round, and if indications are good, a further sum of two or three times that may have to be invested. But the expected reward may be the energy independence of Sweden for a long time to come."

With genuine enthusiasm I thus concluded my 1983 presentation to the Swedish State Power Board. In December 1985 the Swedish Parliament approved the project of drilling to a depth of at least 5 kilometers in the Siljan meteorite crater. The project would be directed by the Swedish State Power Board (a government-controlled energy authority), with additional funding contributed by Swedish investors and the U.S. Gas Research Institute. The Gas Research Institute, based in Chicago, was interested not so much in the commercial success of the enterprise as in the scientific findings that this unconventional drilling would yield.

A spot in the area interior to the Siljan Ring was selected in order to ensure that any oil discovered could not be explained by skeptics as seepage from the thin layer of limestones and sandstones of Paleozoic age surrounding the crater (this layer, in any case, was nowhere deeper than 300 meters). Substantial active oil seepages can be seen in stone quarries within that sedimentary region, but hydrocarbon gas seepages are extensive throughout the impact structure itself, issuing forth from purely igneous rock. Numerous water wells will even support a flame (Figure 6.3).

Drilling commenced in June 1986 and continued until June 1990, when technical problems in the hole made further drilling impossible without a very substantial commitment of additional funds to drill a new branch to the hole. Even so, the results demonstrated that hydrocarbon gases from methane to pentane—as well as light, largely hydrogen-saturated oils—are indeed present deep in the granitic rock.[2]

Four branches of the hole were drilled below 5 kilometers (Figure 6.4), the deepest reaching a vertical depth of approximately 6.7 kilometers. We drilled with a water-based drilling fluid, so as not to contaminate the well with introduced oils, and obtained good measurements of hydrogen, helium, methane, and the other hydrocarbon gases up to pentane (C_5H_{12}). In the variations with depth, there was a clear correlation of all the gases with one another, including helium, a result that

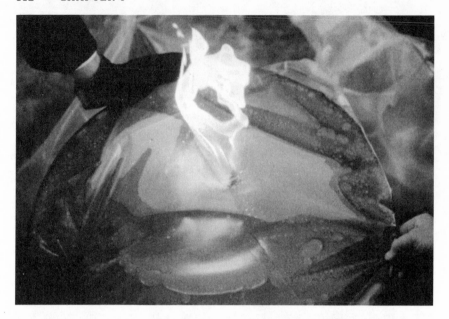

Figure 6.3 Flame supported by gas emissions over a water well in the Siljan Ring. Methane emissions in some places within the Siljan Ring of Sweden are strong enough to produce a flame. For this photograph I covered a water well with a plastic sheet for a few minutes, than pricked the sheet with a pin, and put a match over the hole. A flame shot up 30 to 40 centimeters and then declined to 10 centimeters. I aborted the experiment 10 minutes later when the plastic began to melt. A video I was given while in the region showed a flame 40 to 50 centimeters long emerging from the running faucet in the kitchen of a local farmer's home.

excluded the possibility that they were in any way the result of drilling additives put in from above. In general, the volumes brought up in the returning drilling fluid increased as the depth increased—a strong sign that the hydrocarbon source resides at greater depth. All these results confirmed the abiogenic theory of petroleum formation and supported my view that enormous quantities of hydrocarbons were still streaming up from a primordial source in the deep crust and upper mantle. A most welcome surprise, following upon these results, was our encounter with a huge amount of a concentrated, very fine-grained sub-

Figure 6.4   Site of the drilling project in the Siljan Ring.

stance of great significance—the discovery of which, as sometimes happens in science, occurred only because of a mishap.

Magnetite and Microbial Geology

I n June 1987, just a year after operations commenced, a most remarkable sequence of events occurred. As a result of a drilling mishap, the drill bit was stuck at a depth of 6 kilometers for a period of ten days. During the shut-down, circulation of drilling fluids was not maintained, which made it possible for fluids from the environment to enter the bottom of the pipe from below. When the drill pipe was finally freed and brought to the surface, the lowest 10 meters were tightly blocked by a very stiff paste. Even the high-pressure pumps available on the surface could not blow it out, and the pipe had to be cleaned out mechanically. The material was black, had the consistency of putty, and emitted a strong and objectionable odor.

A sample of the material was sent for analysis to the Norwegian Petrochemical Laboratory, Geolab Nor. The laboratory analysis showed the oil of the black paste to be chiefly a light oil with smaller amounts of heavy molecules whose precise identity was not established. Geolab Nor stated that this oil did not show any resemblance to any of the drilling additives that had been supplied them for comparison.

On a visit to the site three months later, I acquired a sample for myself. Although some 60 kilograms of this thick, black paste had been bored out of the pipes when they were finally brought to the top, nearly all of it had been thrown away, presumably because it was judged an uninteresting, malodorous nuisance of no commercial value. Nevertheless, it was of extraordinary scientific value. Sadly, all that remained in a preserved sample was to be found in one small plastic bag.

The on-site chemist told me that the smell of the clogged pipe when it was drawn up indicated to him that the sludge was some bacterial product and that therefore it must be something that had fallen in from the top. Thus contaminated, the sludge would have had no scientific value. How 60 kilograms of a uniform black paste would have fallen in at the top was not explained, nor was the reason why it would have oil as a binder, rather than water, which was the drilling fluid at the time. All but the small sample in his hand had been bulldozed into some ditch and covered with dirt to get rid of the stench. The small sample he had retained was in an ordinary polyethylene bag. Because polyethylene soaks up and transmits oils, the lighter hydrocarbons would have escaped easily. The result of this inadequate preservation was that the material was now rather stiff and not so pliable as had been described initially.

It so happened that I was to go directly from Sweden to a friend's house for a brief vacation on the Spanish Mediterranean island of Mallorca. I arrived there on a weekend, and for a day and a half I had no access to anything one might buy in a hardware store or a drug store. Yet I was fascinated by the black material and could not wait that long before analyzing it. I thus decided to attempt a little kitchen sink (and kitchen cabinet) chemistry experimentation.

I first looked through the apartment for anything that might serve as an oil solvent, but there was no paint solvent, no nail polish remover, or anything of the kind. There were, however, magnets in the house—the magnetic latches of cabinet doors—which might give me an indication of one of the substance's properties. I unscrewed one and ascertained that the sludge was strongly magnetic. Then I put a small amount of the material in hot water and kitchen detergent, and indeed, after some effort, it dissolved. The dilute liquid so produced was almost transparent; it had just a slight gray, hazy appearance. It evidently contained particles, but they were very small. I put a drop of this fluid on a piece of aluminum foil, and in the great dilution it looked utterly transparent. Then I held the door-latch magnet underneath, and instantly the two lines of the magnetic poles appeared as distinct black lines on top of the foil. Similarly, when I held the magnet to a side of a glass filled with this liquid, it immediately produced a large black patch on the wall of the glass. Further experimentation demonstrated that this material consisted mainly of very small magnetic grains and an oil that could be dissolved in water with kitchen detergent.

I next put a sample of the original paste in the freezer overnight and observed that it was much stiffer after that, but it was not completely hard and could still be squashed and deformed. This implied that it contained extremely little water, which would of course have frozen. Thus, although the drilling system was entirely filled with water at the time the material was found, this sludge must have entered in an oil-based fluid at a sufficient concentration to replace the water and not mix with it at all.

I also poured some drops of this dissolved material onto kitchen paper towels to see whether the towels would serve as a crude chromatograph, by which one sorts out different components of a fluid, depending on the mass and therefore on the speed at which they diffuse in a porous material. A black patch of a definite size was formed, surrounded by a much larger but also clearly defined wet area. This meant that the particle size was small enough to be transported in the paper by diffusion. Furthermore, when I cut a lump of the material with a knife, the cut surface was glossy. I could judge from this that the particle size could not be much larger than the wavelength of light.

So what was all this? The only black and magnetic material that I knew to occur in nature was magnetite (Fe_3O_4). What I seemed to have detected, therefore, was unusually fine-grained magnetite, yet in a size range large enough to be ferromagnetic. Particles of magnetite smaller than about 3×10^{-6} centimeter would not support the cooperative phenomenon of ferromagnetism. The strong odor emanating from the sludge could be the smell of a heavy oil together with rotting biological material. A dead rat on the garage floor would be the best description I could give.

Fine-grained magnetite was surely a major component of the black sludge, but why was it there at all? How had magnetite become concentrated in a hydrocarbon fluid that could move through pores and cracks of granitic rock sufficiently to enter an untended borehole? And what might a dead rat have to do with it?

I had many laboratory analyses done on the sludge, demonstrating the fine-grain nature of the magnetite, which ranged from just fractions of a micron to no more than a few microns in size.[3] Mössbauer spectroscopy performed in two laboratories[4] showed the presence of zinc at the level of about 2 percent, making zinc the second most abundant metal, after iron. Because the zinc was contained in the crystal lattice of the magnetite (and because zinc was not a component of the drilling hardware, fluid, or additives), it was clear that zinc must have been available during the formation of these crystals. I recalled, too, that in the vicinity of the Siljan Ring there had indeed been a commercial zinc and lead mine. These findings confirmed that the sludge was native to deep rock, not a fabrication of drilling fluid injected from the surface.

Neutron activation analysis[5] also revealed a similarly high level of zinc, as well as a number of other anomalies. This investigation compared magnetite grains contained in the sludge with the coarser magnetite (millimeter rather than micron size) contained in the drill cuttings at the same depth. In addition to a size difference, the two sources of magnetite showed numerous other large differences in trace element content, many by factors of more than 10. The quantities of magnetite contained in the sludge were also very much greater than the concentration of magnetite in the surrounding granite. In abundance as well as size and chemistry, the grains of magnetite in the sludge were quite different from those in the rock environment. This indicated to me that the source of the magnetite or its precursor iron molecule resided at a greater depth and that the leached molecules or grains were borne upward by an ascending hydrocarbon fluid.

Another important laboratory finding was the unusually high level of iridium in the magnetite of the sludge, which proved to be 250 times higher than in the larger magnetite grains that were a usual component of the granite and that were selected from granite cuttings derived from the same depth. The investigators stated that these were the highest iridium values they had seen in anything other than oil wells. Not only this black oily material but also oil shales in the shallow and ancient sedimentary rocks surrounding the rings had previously been found to be enriched in iridium. Fluids must have come up from deep in the earth, bringing with them iridium compounds that could then be traced in other oil-soaked minerals.

This conclusion is based on conventional understanding in the geosciences. Iridium is a very heavy metal. During the earth's early phases of differentiation, iridium and other heavy metals (including abundant iron and nickel) migrated toward the interior, becoming the earth's metallic core. Unusually high concentrations of iridium found anywhere in or on the crust could have come from only one of two possible sources. The iridium could have been transported by a fluid upwelling from the depths of earth, or it could have been delivered by a meteorite. The former seems more probable, because there is a strong association of iridium with oil wells.

An additional investigation was initiated by my friend Robert Hefner, a deep-gas entrepreneur. The investigation was carried out by Paul Philp, at the University of Oklahoma. Philp is a specialist in the investigation of biological molecules in petroleum, and he analyzed the oil in the black sludge for such molecules. First, he concluded that the oil was not any form of contaminant introduced by the drilling process or from any of the additives. It was a natural material. Second, he saw that a number of a class of molecules called steranes were the same set, and in much the same proportions, as he had previously detected in the oil seeps at the surface of the Siljan Ring and in oil shale that existed in the shallow sediments surrounding the ring. The view that oil shale in the sediments had been the source material for the liquid oil seeps would be the common assumption in petroleum geology. Yet he now found the same fingerprints in the oil brought up from a depth of 5 kilometers. "How it got down there I do not know," was his response to this strange finding. My response, of course, was that all three types of oil had traveled upward from deep levels.[6]

Philp had also identified one molecule as characteristic of a product of a marine organism, and he thought this to be further proof of a downward migration of the oil produced at shallow depth. However, I later ascertained that the same molecule was a common product of methane-oxidizing bacteria, which are scarce in the surface biosphere but, I believe, abundant at depth.

The most remarkable recent analysis of the Siljan sludge revealed the probable origin of all the magnetite, which would also account for its concentration and for the stench of the sludge. The answer: Life was responsible. I had suspected the fine-grained magnetite to be a bacterial product, even though temperatures at the depths at which the magnetite was found ranged from 60°C to 80°C. Magnetite is one of the substances left behind when more highly oxidized iron is reduced by bacteria. (Magnetite, Fe_3O_4, contains 16 atoms of oxygen for every 12 atoms of iron, whereas ferric iron, Fe_2O_3, contains 18 atoms of oxygen for every 12 atoms of iron, so magnetite is the reduced form.) The microbes, therefore, were scavenging oxygen atoms from ferric iron in order to burn the hydrocarbons that were streaming past them. Magnetite was the by-product of this metabolic activity.

Thus it seemed necessary to make an effort to culture microbes from these depths, and I asked the Swedish National Bacteriological Laboratory in Stockholm whether they would try. Dr. U. Szewzyk expressed great interest, and he and his team decided to make the attempt. They designed a sampling apparatus on a wireline, with many capsules at different depths. The device was introduced at a time when drilling had stopped and water from the formation was filling the hole. The result was strikingly positive.[7] At least two previously unknown strains of bacteria were successfully cultured, both in a temperature range similar to that at the sampling depth, and both in anaerobic conditions also similar to those at the sample locations. Although acetate and sugars were used as the nutrients to support bacterial growth (these substances are commonly used for bacterial cultures), rather than hydrocarbons and iron oxides, the fact that any life was present at all and that magnetite was present in large amount was significant. Because of the character of the nutrients, the cultured microbes were somewhat removed from what was probably the first stage of the food chain; presumably, they fed on the microbes that were nearer to the primary step. Nevertheless, they did demonstrate the presence of at least one node of the ecology in the deep biosphere of the Siljan Ring.[8]

Thermophilic microorganisms were indeed present at depth in the Swedish well, and they can be assumed to have been responsible for the production of the large quantities of magnetite that had invaded the drill pipe. Corresponding to this, large quantities of hydrocarbons must have been used up in the reduction of ferric iron to magnetite, which is the lowest oxidation state of iron that could be reached by reduction with hydrocarbons.

The concentrated magnetite was far more than an isolated curiosity, however. The same sort of fine-grained magnetite was found in abundance in a very similar oil encountered within a second borehole that our Swedish team drilled soon after the first borehole. The second site was at the center of the Siljan Ring, 11 kilometers distant from the first hole. This second find provided strong confirmation that the earlier discovery had not been a local anomaly and had not been produced by any drilling additives (as some critics had claimed), because for this hole, water had been the principal drilling fluid, and no oils remotely similar to crude oil had been allowed near the hole.

The entire Siljan Ring, an area of about 1600 square kilometers, shows a strong positive magnetic anomaly centered on the circular feature. Magnetite was the only magnetic mineral that we had encountered, and one could calculate the quantities that would account for the anomaly. The results were comparable to the quantities in other Swedish magnetite deposits that had long been mined as the best source of iron ore for steel production. This suggests that similar magnetite sludge produced by the same sorts of biological and geological processes apparent at depth in our Siljan boreholes may have produced Sweden's numerous and commercially valuable magnetite iron ore deposits, from which the country's highly successful steel industry was built up. If microbially made magnetite was indeed the origin of all the Swedish magnetite deposits, this would represent a case for microbial geology on a large scale.

The oil sludge drawn from a depth of 6 kilometers in a purely granitic and igneous region of Sweden is compelling evidence of the presence of hydrocarbons at a depth that the biogenic theory cannot account for. The sludge thus provided strong confirmation of the deep-earth gas theory. Culturing experiments, in turn, provided tantalizing clues to the presence of deep microbial life. More than just theoretically useful, the demonstration that unusual concentrations of magnetite are correlated with hydrocarbons has also proved to be of practical importance in petroleum exploration. It is now widely regarded that positive magnetic anomalies, which can readily be located, are indeed indications of the presence of hydrocarbons.[9]

There is one more installment to this story of petroleum discovery in the Siljan region. In April 1990 a downhole pump was installed in the first hole to see what could be pumped up (a common procedure in the petroleum industry). All the samples taken previously had been from fluids and sludge caught in the drilling equipment. This pumping operation pulled up some 12 tons of crude oil, described by the Danish Geological Survey as "looking like ordinary crude oil." Along with the oil came 15 tons of fine-grained magnetite. Concentrations of hydrocarbons were seen in the rock cuttings of the second hole, which were sampled every 5 feet of depth. All the extremely high values came each time from samples where the drill had crossed a volcanic intrusive rock, dolerite, which is a known feature in the granite of the area. The intrusive volcanic rock had certainly come up from deeper levels, and this suggested that these intrusions were the conduit for the hydrocarbons. I could not have hoped for a stronger confirmation of the deep origin of the hydrocarbon fluids (Figure 6.5).

Thanks to the 1990 pumping results, the quantities of oil and magnetite paste found at depth could no longer be dismissed as "trace amounts," as they had been described earlier in several scientific journals. Nevertheless, no major journal would publish these striking results, and I received responses from referees that called these observations totally incredible and maintained that they would have to be repeated by another team before they could be accepted for publication. Our invitations to major petroleum research organizations to send delegates to the site and observe our actions went unheeded.

Eighty-four barrels of oil are meaningful, especially when they are found in a location where, in the conventional view, not a single drop of oil could have a rational explanation. The theory of the abiogenic origin of petroleum had thus been confirmed. Dr. Peter N. Kropotkin, a distinguished petroleum geologist in the former Soviet Union wrote, in an issue of *The History of Science,* "The discovery of oil, deep in the Baltic Shield, may be considered a decisive factor in the hundred-year-

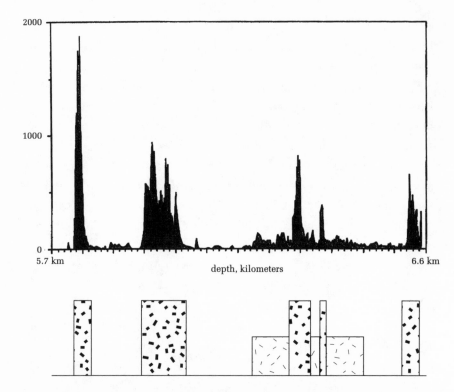

Figure 6.5 Methane concentrations in rock cutting grains between depth of 5.7 km and 6.6 km. Upper figure shows methane extracted from interior of the grains. A consistent extraction procedure was used, so the relative values are correct, even though the absolute values cannot be defined accurately. The lower figure indicates, on the same depth scale as above, the occurrence of volcanic intrusive rock, dolerite. The high columns represent pure dolerite grains, the low ones represent a mixture of granite and dolerite grains. The rest of the line represents pure granite grains. The correspondence of high values of methane with dolerite intrusions is evident.

old debate about the biogenic or abiogenic origin of oil. This discovery was made in deep wells that were drilled in the central part of the crystalline Baltic Shield, on the initiative of T. Gold."[10]

Nevertheless, no commercial flow rate could be established in either hole drilled in the Siljan region. That was not, however, because the supply was meager. Rather, a short time after a flow had started, magnetite paste always blocked the cracks in the rock entering the wellbore. (Had the technical resources, the money, and the interest been available to drill another, substantially deeper branch of the hole, the ultimate source of the hydrocarbons might well have become available in commercial quantities. But further drilling would have been a gamble for the investors, so they decided to call it quits at that point. As it stands, and like many drilling ventures in the petroleum business, the project never proved to be commercial, but it was nonetheless a scientific success.)

What a cruel and ironic turn of fate! *Commercial* success was not possible because of the great abundance of the very substance that bestowed on this project its *scientific* success. Magnetite sludge made sustained petroleum production impossible, but it lent important support to the deep hot biosphere theory.

In the Swedish drilling project, I had been granted a glimpse of the deep hot biosphere. I now thought it quite possible that subsurface microbiology was so widespread that every oil-bearing region had been subjected to biological alteration, down to the deepest wells from which oils have been extracted. Because the earth's temperature increases with depth, the microbial life forms involved must be hyperthermophilic, living at temperatures up to 120°C, possibly as high as 150°C. And as explained in Chapter 5, I soon came to suspect that the quantity of these life forms, in terms of mass or volume, could be at least comparable to the quantity of all the surface life we know. The deep hot biosphere theory would solve the paradox of seemingly conflicting facts that had long split petroleum geology into two camps and had stalled reconsideration of the origin of petroleum for many decades. Might this new view of life *within* the earth call for a rethinking of much of the rest of geology, as well?

Chapter 7 Extending the Theory

‎

O nce we come to understand the existence and immensity of hydrocarbon sources streaming up from the earth's mantle, we can profitably revisit a number of subjects in geology. Some of these subjects have long been regarded as geological puzzles; others have seemed settled but, I believe, require a fresh look.

In this chapter, I will explore two extensions of the deep-earth gas theory. The first is an interpretation of the genesis of diamonds at great depth within the earth. In the second, I present a new view of how some metal deposits have arisen and become concentrated in clusters in the outer crust. Because both topics are persistent problems in geology, the speculations offered here might pique some special interest. Chapter 8 will then pursue a third, and very controversial, extension in the geosciences of deep-earth gas theory: an explanation of the earthquake process.

Chemistry at great depth is likely to be quite different from the low-pressure chemistry with which we are familiar. At a depth of 150 kilometers, for example, the pressure would be 40 kilobar, which is equivalent to 40,000 times our atmospheric pressure. Very many different molecules will be held together by that level of external pressure—molecules that we have never seen at the surface. In fact, the very concept of molecules begins to break down at pressures comparable to the forces within a molecule that hold it together or make it come apart. We may have seen at the surface some of the degraded products of these unknown molecules or crystallographic forms, but we cannot produce them or examine the antecedents. Nevertheless, they may play a major role at depth.

Conducting experiments at a pressure of 40 kilobar or more, and at the elevated temperatures that occur at depth, would require extremely expensive apparatus. Furthermore, thermodynamic calculations for understanding the secrets of deep chemistry are complex and very difficult to perform in an exploratory way. If one knows what one is looking for, some information may be obtainable. But as a means of determining unknown molecules, calculations by themselves are not very suitable. The circumstances in which the atoms of molecules derived from this largely unknown chemistry are found may shed some light on the chemical processes that were involved, especially because there are many clear regional correlations in the deposits of certain metal minerals and, in the case of carbon deposits, associations with features in the crust.

What are the processes in the earth that have concentrated certain materials in well-defined locations in the crust? One might have thought that the tendencies would go the other way—that subsurface earth processes would arbitrarily mix things up. But then why could we ever pick up a nugget of gold or a diamond made from very pure carbon? Or why would we find locations where some particular metals are concentrated in a rock by factors of a million or more relative to other rocks?

Some powerful processes of concentration must be working in the earth, driven by internal energy that the earth possesses. One such source of energy is the gravitational field, which would tend to make heavy substances sink and light substances rise. We can understand the formation of an iron core in that way, because iron is abundant and about twice as dense as the rocks. It is equally easy to understand the arrival at the surface of water and other fluids that are less dense than most rocks. But there are many concentration processes that are defined by the chemical properties of a substance, not just by its density. It is generally believed that all those cases must involve a liquid flowing through pores in the rock, and its chemical properties must be such as to pick up from the rock, and carry along in the flow, the specific atom or molecule that is to be concentrated. Then some change of circumstances—such as pressure, temperature, or a chemical agent—must decrease the quantity of the substance that the fluid can transport, thus leading to a deposition of the substance, now highly concentrated compared to its initial presence in the rocks. All sorting processes require energy, and that energy must be coming, at least in part, from chemical energy that the earth has possessed since its formation.

Although many observed concentrations of chemical components have been satisfactorily explained by reference to such processes, many others have long presented major puzzles for geology. There is nothing more interesting or more important in science than the observations that we cannot explain. The formation of diamonds and the laying down of certain metal ores belong to this category. Let us turn to these two puzzles now.

The Origin of Diamonds

The discovery on and near the earth's surface of crystals of pure carbon—diamonds—was wholly unexpected. These are not crystals that are stable and in equilibrium at low pressures, and for this reason diamonds could not have formed anywhere near where they were found. Even if a near-surface process could concentrate carbon to high purity, this should have led to the deposition of graphite, the stable crystallographic form of carbon in the earth's crust. Diamond is the high-pressure form of carbon, but the pressure required to reach this stable condition is so immense that one would hardly have expected to find samples at the surface. Not only could they not have formed here, but they will in fact decay to graphite (the black stuff in your pencil) in the course of time. Diamonds are not forever, but as it turns out, they are for long enough.

The pressure needed to reach the domain of diamond stability is approximately 40 kilobar, or 40,000 times our atmospheric pressure.

This is known from theoretical calculations and has been confirmed in high-pressure experimentation. We do not find in nature any pressure vessels that could have withstood such pressures. The only locations we know of in which we would expect such pressures would be at a depth at which the overburden weight of rock would balance that pressure. The earth is not built like a steam boiler, where the tensile strength of steel will contain high pressure; rather, it is built like a pile of rubble, arbitrarily thrown together and possessing no tensile strength. This lack of tensile strength enables us to calculate the minimum depth at which the pressure required for diamond formation could be reached, and that is approximately 150 kilometers—and thus in the mantle well below the crust. This leaves two things to be explained: how diamonds came to the surface and why they retained the high-pressure form at the surface and did not degrade into graphite.

The essential discovery that answered both of these questions was made near a town called Kimberley, in South Africa, in 1870. Diamonds were found there, but so too was an extraordinary feature: a steep, funnel-shaped depression that went deep into the rock, narrowing from about 200 meters at the top to a few meters at a depth of about 1 kilometer, and continuing downward as a pipe to beyond the depth of observation. By now 10 such funnel features bearing diamonds have been found, distributed around the globe. They contain some mantle rocks, together with local fill that has fallen in. This fill and the ground nearby display a high concentration not only of diamonds but also of a rock called kimberlite, which appears to have come from great depth. (A few more such features have been found containing kimberlite but no diamonds.) Although these pipes are often called volcanic, no evidence of frozen lava has been found in them.

An extraordinary picture thus emerges: Diamonds must represent enormous gas blowouts, presumably from a diamond-forming depth such as 150 kilometers. Enough fluid pressure must have built up there to blow a hole through all of the overlying 150 kilometers of rock, and the erupting gases carried up material from the great depth. This, then, is how we come to have natural diamond at the surface. But the eruption process also explains why the diamond persisted and did not decay into graphite. In this fast eruption the driving gas would cool rapidly, and with it the diamonds. Then, at the low temperature of the surface, the diamonds would no longer have the internal energy to convert their crystal configuration and so would remain an unstable solid in a "supercooled" state. (We all have some acquaintance with supercooled configurations, which include steel knives that have been heated and then cooled rapidly and thereby turned into a harder and more brittle steel).

Why are diamonds so rare at the surface? The concentration of pure (unoxidized) carbon at diamond-forming depth and the occurrence of a blowout from there all the way to the surface are both improbable. Which of these is the factor that limits the quantities delivered? Is there a connection between them, such that the places in which carbon became concentrated also provided large bubbles of high-pressure gas in the rocks? Or are the two phenomena independent of each other, in which case a very high content of diamonds in the deep rocks would be presumed, so that each deep blowout had a good chance of bringing up precious stones? We do not yet know the answer to this tantalizing question. Worse still, we cannot even suggest a line of investigation that could be pursued, within the limits of our present ability and knowledge.

What happens to diamond-bearing rock, however, when geological processes force it to the surface at a slow speed? A slab of rock in North Africa is composed of materials known to indicate a very deep origin. This rock contains many inclusions filled with graphite, which is the crystalline form of pure carbon at low pressure. But that graphite reveals the octahedral shapes characteristic of the crystallographic structure of diamond. These inclusions started as diamonds, it has been plausibly argued, but in the course of a slow ascent and gradual cooling, the carbon atoms reassembled into the low-pressure form.[1] This is an important observation, for it indicates that at least in some areas, diamonds were very abundant at their formation depth. The African slab of graphite inclusions suggests an original abundance of diamonds more than 100,000 times greater than in the kimberlite pipes. This in turn implies that carbon-bearing fluids were abundant at those levels also and that they deposited clean carbon. Perhaps diamonds are rare at the earth's surface not because they are rare at depth but because the eruptions that can transport them quickly are rare events.

Diamonds often contain inclusions of high-pressure, carbon-bearing fluids, notably methane and carbon dioxide.[2] One of these fluids is presumably the principal carbon source from which pure-carbon crystals originated at depth. I tend to believe that methane or the other light hydrocarbon gases were principally responsible, because they are more easily dissociated into component atoms than is carbon dioxide. By far the dominant quantity of gases enclosed within diamond is nitrogen. We do not know the reason for this association, but it could be that nitrogen will lead to the formation of ammonia (NH_3) and thereby rob methane of its hydrogen, causing the deposition of carbon. As yet we do not know the high-pressure chemical equilibrium between these two substances; all we know is that they are often seen together on planetary bodies and that neither destroys the other at low pressure.

The carbon isotopic ratios in diamonds have been used in a study of diamonds originating in two different types of rock. The authors reach the conclusion that neither the influence of recycled biogenic carbon nor the global and primordial heterogeneity of mantle carbon are likely for the origin of the large C-13 range; the data instead support a fractionation process.[3]

Hydrocarbon fluids present a problem similar to that of diamonds, but their abundance in the earth's upper crust blinds us to their anomalous presence in our realm. At high pressures, hydrocarbons represent the stable configuration of hydrogen and carbon. Hydrocarbons should therefore form spontaneously in the upper mantle and deep crust. But at low pressures at or near the earth's surface, liquid hydrocarbons are supercooled, unstable fluids. As they upwell into lower-pressure regimes, they begin to dissociate, and this means they begin to shed hydrogen. This is exactly what we see in the vertically stacked patterns of a hydrocarbon region that go from methane at the deepest levels to oils and eventually to black coals at the shallowest levels. Each step in that stack is one of further hydrogen loss.

Overall, the abiogenic theory of petroleum formation offers the possibility of a more thorough explanation for the genesis of diamonds. I have raised objections in earlier chapters to several widely held beliefs in modern geology, such as "The existence of pore spaces at depths of more than a few kilometers is highly improbable"; "Unoxidized carbon cannot exist at levels deeper than the sediments"; "Hydrocarbons are not stable below a depth of about 15 kilometers"; and "Gas emission from deep levels (where there are no pore spaces) cannot occur." The diamond formation process I have described is in sharp disagreement with each of those beliefs. The existence of diamonds thus tells us that there are pore spaces at mantle depths and that they can be filled with carbon-bearing fluids; that these pore spaces allow fluid flow through them; that unoxidized carbon (the diamonds themselves and hydrocarbon gases) can and do exist at these deep levels; and that gigantic gas eruptions from these levels can occur. I do not know of any other process that could bring together clean carbon to make centimeter-sized pieces of diamond, nor any expulsion process other than the one I have described. With the knowledge that the diamonds have provided for us, we can next assess whether the abiogenic theory provides an opportunity to understand better the mechanism by which other important resources—concentrated deposits of various metals, including copper, iron, zinc, lead, and uranium—have come into being.

A New Explanation for Concentrated Metal Deposits

In places throughout the world, and especially in northern South America and in the Wyoming–Montana region of the United States, metals may be found in concentrated clusters of deposits that may include copper, lead, zinc, silver, and gold in close proximity. How do all these kinds of metals come to be found in the same neighborhood, and each in concentrated deposits? What processes could single out an element and cause its deposition at a concentration a million or even a hundred million times greater than in the source composite from which it came? We find high concentrations of certain metals even in rocks like granite where such deposits clearly represent intrusions that occurred after the magma solidified.

Several general conditions must be satisfied for metal concentrations to form. First, as already discussed, there must be a fluid that can flow through pore spaces and fissures in a source rock where the metal is only sparsely distributed. This fluid must be able to gather up—that is, to *leach*—the metal from the rock and carry it along with the flow. Leaching is a very energy-intensive process. So in order to perform this feat, there must be a source of physical pumping energy that can push the fluid through the rock. Then, to make an ore deposit out of the leached and transported material, the fluid must encounter conditions that cause its metal cargo to drop out of solution. Those conditions may include a decline in temperature as the fluid rises toward the surface. Or perhaps mixing with and contamination by a different kind of fluid will change the chemistry and solubility to a point where the metal drops out of solution. Possibly, too, a threshold loss of pressure during the upward journey might trigger the dissociation. And, as we shall see, life in the deep hot biosphere may even play a supporting role.

Hot water is generally considered to be the fluid responsible for creating concentrated metal deposits, but the hydrothermal theory cannot account for realistic processes that could concentrate some of the metals. Indeed, the problem is so great that answers are promoted piecemeal—some chemical reactions are proposed for the solution and deposition of one metal, and a different set is proposed for another. Piecemeal answers are especially questionable when there is a group of metals involved, and a different path is suggested for the formation of each of them, yet they are often found closely packed together. The problem is more general, and so one solution should be found that adequately explains the collective phenomena. There are groups of metals whose shallow deposits are often found in close association, such as zinc with lead, and gold with silver and other heavy metals.

There are two great difficulties with the hydrothermal theory of the formation of metal ores. First, many metals, especially the heavy metals, are not sufficiently soluble in water at any temperature—even in brines hosting aggressive salts. Geochemist Konrad Krauskopf, for example, noted in his textbook that the insolubility of many metals and their compounds in water is "a long-standing difficulty with the classical hydrothermal hypothesis of metal deposition." He gave examples of the enormous quantities of water, bearing mere traces of metals, that would have to have flowed through cracks in order to account for the accumulation of known metal deposits. He concluded that those quantities were quite unrealistic.[4]

The second problem with supposing a water-based fluid for the deposition of metals is a likely deficiency in pumping power for leaching metals in the first place. Water circulates through the outer crust of the earth, but only rarely (if at all) to depths of 10 kilometers. Boreholes that go to such depths have a much higher probability of turning up light hydrocarbon fluids than water. If we assume, therefore, that water-facilitated leaching takes place at a depth of no more than about 10 kilometers, then the pressure driving the pumping action at such depth would be given at most by the overburden weight of rocks, including the weight of its contained fluid (which is usually a very small fraction of the total weight). The maximum power available to drive the flow and hence the leaching, on the assumption that water and rock were initially in pressure equilibrium at depth, would be the volume that is expelled per unit time, multiplied by the pressure difference between the entry and exit points (leaving out the component that is derived just from the static head of water). At a depth of 10 kilometers the rock overburden weight would create a pressure of about 3000 bar (3000 atmospheres), and the static head of water would be about 1000 bar. Thus 2000 bar (a mere 2 kilobar) would be available to drive the fluid through the rocks. This is not an impressive amount of pressure for the task required, and it would not give the fluid a strong leaching ability.

Better candidate fluids for the leaching and transport of metals are hydrocarbons. Hydrocarbon fluids surpass water in both the capacity to hold metals in solution and the pumping power required for the energy-intensive leaching process. Consider, for example, the leaching power of a hydrocarbon fluid that flows upward from a source depth around that at which diamonds formed, undoubtedly by the dissociation of clean carbon from a carbon-bearing fluid. At a depth of 150 kilometers, the pressure bath would be 40 kilobar, or about 40,000 times the atmospheric pressure at sea level. Using the same density assumptions as in our previous example, we would expect a driving force for leaching of about 35 kilobar. This is far greater than the 2 kilobar of leaching power we calculated in our previous water example. Even by itself, this fact should alert us to the need to investigate whether these fluids, so highly powered for leaching through great distances of rocks, have anything to do with the metal concentrations that have been deposited from some sort of leaching solution.

Another point in favor of hydrocarbons as the fluid carrier of metals is that we know that many metals are indeed carried in petroleum. Hydrocarbons may enter into molecular arrangements with metals to form complexes called *organometallics*. Organometallic molecules have been identified in every crude oil that has been analyzed; vanadium and nickel porphyrins are the most prominent, but there are several others. Gold and silver organometallic molecules, for example, are detectable in some crude oils, though only in trace amounts.

Porphyrins are a group of organometallics that contain nitrogen in addition to carbon and hydrogen. Porphyrins such as hemoglobin (with a single iron atom at its core) and chlorophyll (with a single magnesium atom at its core) are valuable catalysts and are manufactured by many familiar forms of surface life. The origin of porphyrins found in petroleum was therefore readily attributed to biological debris. That explanation would lead us to expect to see principally magnesium and iron porphyrins in petroleum. Yet not a single case is known of their presence in petroleum.[5] Instead, only nickel and vanadium porphyrins have consistently been found. It seems extremely improbable that on every occasion, in all oils, the original metal atoms were exchanged for just nickel and vanadium from the rocks in their surroundings. Furthermore, it has not been explained how plant debris would have produced the nickel and vanadium molecules when subjected to the relatively low pressures and temperatures that the prevailing view-

point deems necessary for the genesis and stability of biogenic hydrocarbons. In the picture of the deep earth set forth in this book, however,
nickel and vanadium complexes may well be expected to form at the
high temperatures and pressures at depth. Possibly these two are simply the organometallic compounds that survive longest, after many
others have disintegrated at deeper levels beyond our reach.

To the best of my knowledge, no one has yet developed any idea of
the organometallic chemistry that would prevail at a pressure of, say,
50 kilobar. We have, as yet, few laboratory simulations of this kind.
Surely, many molecules would be held together at this pressure, even
though they would readily disintegrate at the pressures near the surface of the crust. Among those, organometallics may be plentiful in
cases where the formation of organometallic molecules occupies less
volume than the materials from which they are derived. Perhaps a
whole range of organometallics are produced by hydrocarbons leaching through rocks at great depths and over vast distances.

Organometallics are not only *built* from hydrocarbons together
with the metals that hydrocarbons can leach from rock on their long
upward journeys; they can also be *transported* by hydrocarbons. Most
organometallics are soluble in hydrocarbon oils and thus will be carried along with the flow. When temperature, pressure, or other solubility conditions reach a threshold at which a particular kind of
organometallic can no longer be carried by the stream, a concentrated
metal deposit would be generated in that spot.

The hydrocarbon flow, on the way up, would generate a large array
of molecules, the particular ones depending on such things as the
carbon–hydrogen ratio at formation, the ratio to other elements such as
nitrogen and oxygen, the catalytic action of specific minerals in the
rocks, and the pressure–temperature regime encountered along the
way. Among those molecules may be one class that is unusually favorable for forming a particular organometallic compound with one metal,
another class with another. The great diversity of hydrocarbon molecules (differentiated by the number of carbon atoms bonded together in
chains, in rings, or in some combination of the two) could thus be the
reason for the selectivity in metal deposits. Different kinds or groups of

metals would occur in closely packed locations, suggesting a general hydrocarbon upwelling in the area, yet the deposits may be segregated because of differences in metal affinities with various hydrocarbon molecules. Hydrocarbons of greater or lesser carbon number or some other structural feature would also be expected to unburden themselves of their metallic constituents at different threshold changes of temperature, pressure, acidity, and solutes. This could well account for the observation that lead and zinc are typically found together, gold with silver, and so forth.

Empirical support for the hydrocarbon theory of metal deposition includes the close association of carbon with gold that is well recorded in both scientific and popular accounts of mineral prospecting. Gold miners in Colorado, California, the Yukon, and South Africa were well aware of this typical association and searched for a "black leader"—a trail of carbon. They would then dig along the blackened passages, with a reasonable hope of reaching a body of rock that contained a useful admixture of gold. Conventional geological wisdom gives no hint of an explanation for this association, but the deep-earth gas theory surely does. Gold has been leached out of deep rocks and transported as an organometallic by an upwelling stream of hydrocarbons. Because of changes in pressure and other conditions along the way, at some point the metal dissociates from the hydrocarbon molecule. And as with coal deposits, eventually the hydrogen, too, escapes from the carrier molecule, leaving behind carbon, or soot, which might then be carried some distance by flowing water—hence the "black leader."

It is interesting that the other substance commonly associated with gold is silicon dioxide—quartz. Silicon is in the same column and below carbon in Mendeleyev's table of the elements (the periodic table), and the two have very similar properties. Silicon is, however, much more reactive than carbon. It is found only in oxidized form, whereas carbon is found in both oxidized and reduced forms. Silicon will form oils that are quite similar to hydrocarbon oils but that sometimes have higher thresholds of thermal stability. Silicon oils and hydrocarbon oils are almost certainly soluble in one another. I do not know (and possibly no one knows) whether silicon–metallic compounds, analogous to organometallics, will form at high temperatures and pressures. An argument in favor of silicon oils as metal carriers is the occurrence of gold in quartz veins rather than in quartz deposits, which suggests a common migration path for both the silicon dioxide and the gold.

Sometimes it is not just leftover soot but hydrocarbons themselves that are associated with metal deposits. For example, the ancient tin mines in the granitic rock of Cornwall, England—mines that supplied ancient Rome with tin—have oil dripping down on the miners as they work. Other sites of metal mining that are associated with hydrocarbons include Wyoming, Alaska, and the Ural Mountains. Methane explosions in the granite of an iron mine in Newfoundland have stopped mining operations there. Despite substantial evidence of the association between metals and hydrocarbons, scientific papers that report these findings still lean toward hydrothermal explanations for the metals, assuming that the hydrocarbons are a mere contaminant, which they ascribe to nearby biological debris.[6]

Many metals will readily form metal sulfides if sulfur is available. Mercury, as the sulfide cinnabar, is often found together with oil and tar. Mercury may come up in a gas stream as mercury vapor or as dimethyl-mercury, having enough sulfur to transmute into cinnabar. This mechanism would also apply to many other metals, which would not resist being wedded to sulfur and thus being transformed into metal sulfides. For mercury, it is particularly clear that the metal has come from great depths; it is strongly associated with helium, in particular with helium high in the isotope helium-3, which is the marker for primordial helium caught in the formation of the earth and making a small addition to the helium derived from the radioactivity of uranium and thorium.

Is the deep hot biosphere involved in any way in metal deposits? One might speculate that microbial activity plays a role in some cases, especially in the formation of nuggets of pure metals. It is quite conceivable that microbes find the hydrocarbon component of the organometallics to their liking and thus take the initiative to strip the metal of its surrounding hydrocarbon at a depth below the threshold at which dissociation would occur abiotically. One may also suspect that different strains of microbial life prefer different organometallic compounds to disintegrate. That could perhaps account for or contribute to the effect that is clearly of major consequence in this puzzle: the clustering of several different metal deposits in closely neighboring regions. These would be regions in which hydrocarbon upwelling was generally strong and in which some microbes became dominant in some patches, other microbes in others.

It is, after all, well documented that familiar surface bacteria are remarkable geochemical engineers, reworking their surroundings in ways that result in the formation of mineral crystals or even large, uniform deposits. It is generally believed that microbes can build concentrated deposits of a wide variety of minerals.[7] For example, the bacterium *Desulfovibria* produces crystals of pyrite (FeS_2), greigite (Fe_3S_4), sphalerite (ZnS), and galena (PbS). Several kinds of bacteria produce magnetite—which, you may recall, was found at the bottom of both of our Siljan boreholes.

Metals enrichment in sedimentary strata is easily explained by scenarios of microbial mediation that remain within the surface-life paradigm. One would simply assume, as most geologists do, that microbes did their work as part of the surface biosphere, while the sediments were first accumulating in a river delta, the bottom of a lake, and so forth. But metal concentrations within veins suffusing igneous rock resist any such explanation. These circumstances support a microbial explanation only if it is accepted that microbes feed on hydrocarbons at depth—that is, only if the deep hot biosphere theory is accepted.

Investigation of these and other possible biogenic avenues of metal deposition has so far been severely limited by the firm and widely held belief that oils could have been produced only from biological materials generated at the surface of the earth and then buried. Biomineralizations that have been studied thus far are almost exclusively concerned with the products of surface life or of those bacterial members of the surface biosphere that make their living in oxygen-poor sediments just below the surface.

Regardless of whether the booty of interest to mining companies is the work of deep-earth hydrocarbons alone or in collaboration with microbial efforts in the deep hot biosphere, I believe that hydrocarbons are indeed the fluids responsible for leaching, binding, and transporting many of the metals. Because of their depth of origin, hydrocarbons offer the motive power needed for extensive leaching, and they may carry many types of molecules so formed in solution. The major uncertainty concerns the formation of various organometallic, high-pressure compounds, and this issue cannot be resolved with the information now available. Nevertheless I believe that hydrocarbons are the best fluids for the pumping, the leaching, and the solubility requirements for moving metals upward through the crust.

Is this all just fantasy? How realistic can these speculations be, ranging as they do into important issues of high-pressure chemistry that are as yet unexplored? Experimentation at 40-kilobar pressure or greater is very difficult or very expensive. Such experiments will not be performed until we find a good reason for undertaking them. Because the preponderant view in the West is that hydrocarbons simply do not exist at depth, there has seemed to be no reason to consider, much less test, the idea that hydrocarbons upwelling from great depths are the cause of concentrated metal deposits. And this despite the strong suggestion of a regional association of clusters of different metal ores with petroleum. It is my hope that the indications and proposed explanations given here will kindle a spark of interest in reexamining entrenched but perhaps invalid assumptions that may otherwise delay progress in understanding the genesis and location of important metal deposits.

Chapter 8　Rethinking Earthquakes

—————————————————————

Earthquakes tell us about the violence that exists in the interior of the earth. Very violent and rapid events obviously occur, but the reasons for these events are not yet fully understood, and many apparent anomalies remain unexplained. I believe that the deep-earth gas theory can go a long way toward developing a new, more comprehensive, and more useful theory of this phenomenon.

According to the deep-earth gas theory, the earth is continuously expelling fluids from great depths, including juvenile volatiles issuing from the mantle. Some of these fluids ascend as part of a stream of liquid rock—magma—that rushes to the surface during volcanic events. Others breach the surface in the more continuous and sedate—but still highly visible—fashion of mud volcanoes, which throw out mud instead of lava. Mud volcanoes may be found in places of ongoing volcanic activity (Iceland), in geothermal areas with no present active volcanism (Yellowstone), and also in relatively cool geological provinces that support commercial oil and gas production (southern Alaska, the oil-rich zone of the Middle East, and the entire Indonesian arc).

Mud Volcanoes

S everal types of geological features on land surfaces, on ocean floors, and on ice fields indicate that an emission of gases from the ground has taken place. These features are often found in regions in which earthquakes are common. The largest of these are the lava volcanoes that not only transport liquid rock to the surface but also open channels from deep levels through which gases can ascend. We know this from the immense and devastating explosions that accompany some eruptions.

A less well-known but still important land-surface feature that is considered indicative of gas eruptions are the mud volcanoes. Here also, gases rise into the atmosphere from the ground, sometimes so explosively that they carry up any soft alluvium (mud) they encounter on their way. The eruption initially is one of gas and mud; the gas disappears into the atmosphere, but the mud settles around the original orifice and dries up, building a mountain that looks much like a lava volcano but lacks the heat of lava. In some large mud-volcano fields, individual volcanoes rise to a height of several hundred meters and develop orifices 100 meters or more in diameter. The base of such mountains may measure several kilometers in diameter. The emerging gases are usually flammable, containing predominantly methane, and in large eruptions they catch fire spontaneously, presumably through electric sparks caused by friction. A photograph taken in Baku (Azerbaijan, a major region for large mud volcanoes) shows a flame 2 kilometers high standing above an orifice 120 meters across.[1]

The quantities of gas that must have emerged to create these giant structures have been estimated, using instrument data for the ratios of gas to mud observed in several eruptions. The figures so obtained far exceed the gas content of the world's largest commercial gas fields.

If mud volcanoes are the gas emission points on land surfaces, we must expect such points to exist also on the ocean floors and in the large ice fields of polar regions and high mountains. How do they appear in these settings?

On the ocean floors, such features have been identified in sonar investigations as circular markings in the ocean mud. These features have been given the name pockmarks and very large fields of densely clustered pockmarks have been identified.[2] The individual circles may be as small as 1 meter across or as large as 200 meters. In several instances (such as in the North Sea), they overlie fairly accurately fields of commercial gas production, and they show enhanced concentrations of methane in the water above them. Pockmarks also show depositions of crack-filling carbonate cements, phenomenon to which I have referred already.[3] It is thought that these markings arise when sudden expulsions of methane gas lift up a quantity of ocean mud, which then settles back in a regular fashion on the floor, thus leaving circular patterns. Sonar can penetrate through some meters of ocean mud and thus reveal levels at which similar fields of pockmarks have become buried by later depositions of mud. These fields are generally at a well-defined interval of depth, which suggests that the emission process occurred episodically, with separate events some hundreds or thousands of years apart. Because massive gas emissions are indicated, they probably coincided with earthquakes.

In permafrost regions of the high latitudes, these gas emission features exist in the form of "ice volcanoes" or pingos. As with pockmarks, pingos also show that episodic emissions of gases have brought up from deeper levels liquid water, which then froze while flowing down the flanks of the volcano. Because ice does not have permanent rigidity but rather flows slowly, these features soon disappear, and only very young pingos can be identified.

There is good reason to investigate whether the emission of gases in mud volcanoes, ocean pockmarks, and pingos have some relation to earthquakes in general, either as an effect initiated by earthquakes or as a cause of them.

A Challenge to Earthquake Theory

The bulk of fluids from the deep earth probably wend their way upward at a pace more leisurely than that of the gases supplying mud volcanoes. En route they create fractures mainly in the upward direction, as the rock overburden diminishes and becomes less resistant to gas-pressure fracturing. (The gas is less dense than the rock and is therefore buoyant relative to it.) Such fractures will then serve as conduits through the solid lithosphere. By the time these fluids reach near-surface pressures, most will have become invisible gases: methane, carbon dioxide, hydrogen sulfide, and perhaps hydrogen, as well as nitrogen, helium, and various trace gases such as radon. Seeps of crude oil represent the visible, liquid fraction of upwelling fluids from the deep.

The prevailing view is that earthquakes are caused by the catastrophic release of tectonic forces that gradually build stress in the rocks. At some threshold value, the strength of the material is exceeded, and a sudden shift in the rocks takes place, producing an earthquake. Some fluids may be emitted as a consequence of mechanically caused fractures of the rock, but in that theory, they play an entirely passive role.

I think that upwelling fluids from deep in the earth, from regions of greater pressure than that exerted by the rock overburden weight, will have several earthquake-related effects. Specifically, they will create fractures and thereby change the previous stress pattern sometimes produced by the forces of unknown origin that are held responsible for plate tectonic movements. A sudden influx of gas from below will suddenly weaken the rock by creating new fractures, and will bring it to the breaking point even under the previously imposed stress. An inflow of gas would also expand into the fault lines and, by holding the faces apart, would greatly reduce the internal friction, facilitating earthquakes in that manner, too.

After the puff of fluid passes into the atmosphere, the pore spaces that had been created in transit may collapse; such a collapse offers a sound explanation for the vertical displacement of chunks of crust during earthquakes and for the volumetric changes in sea floor or continental shelf that would be needed to induce tsunamis. In the great Alaskan earthquake of March 28, 1964, for example, some stretches of land sank within seconds by as much as 30 feet. Presumably this means that the ground below suddenly became denser. But rocks are not compressible to such an extent, nor would such compression occur suddenly. Pore spaces that had expanded the rock with high-pressure gas must have been involved, and when the gas abruptly found an escape route, the pores collapsed. No fluid other than a gas could have supported the rock and then got out of the way in seconds. Similar events have been recorded in many historical earthquakes.

The present viewpoint, popular in Western countries, is that earthquakes are of purely tectonic origin, caused by an increase of stresses in the rock. But this viewpoint came into being only around the start of the twentieth century. At about that time, the seismograph was invented and put into use.[4] Its availability meant that earthquakes could be investigated in fine detail from the seismic records obtained. Data could now be collected by seismographs installed in places far removed from an earthquake event, and those data would be utterly quantitative and untainted by subjective interpretation. The invention of the seismograph meant that it was no longer necessary to experience an earthquake directly, or to interview someone who had, in order to assemble data on the event. The opportunities offered by this new technology were rarely supplemented by eyewitness reports. Such reports, which were inevitably qualitative and tarnished by subjectivity, unfortunately were no longer believed to hold any value for the scientific venture. But there is much that can still be learned from them.

Eyewitness Accounts

When eyewitness accounts are reported today, they are very similar to those that were gathered and recorded over the course of many centuries. The similarity of reports far removed in geography and time confirms their veracity. Eruptions, flames, noises, odors, asphyxiation, fountains of water and mud—all these are recurrent themes today, just as they were in antiquity. The earthquakes surely did not change their character. Only the investigators shifted their attention.

My colleague Dr. Steven Soter has collected historical written records published in various countries of eyewitness accounts of phenomena associated with major earthquakes from antiquity to modern times.[5] Here I shall present a smattering of some of the more interesting, but nevertheless wholly representative, examples. These will help to illuminate why earthquakes can best be understood as outgassing phenomena, within the framework of the deep-earth gas theory.

Earthquakes in Greece and Italy are fairly common, and many references to them are made in the classical literature of Greece and Rome. At that time, volcanoes and earthquakes were the only sources of information about the deeper ground of the earth. What was down there was imagined to be rather terrifying, and for this reason alone, these phenomena attracted a lot of attention.

Aristotle, whose classical writings and authority dominated explanations of natural events in the West for 1800 years—sometimes with correct and sometimes with incorrect theories—provided the first detailed discussion of the earthquake process. According to him, the theory that gases ("air") were responsible for earthquakes was first proposed by Anaxagoras, who said that "the air, whose natural motion is upward, causes earthquakes when it is trapped in hollows beneath the earth." In A.D. 63, Seneca wrote a review of the earthquake literature of the time, stressing that "It is a favorite theory of most of the greatest authorities that moving air is the cause of earthquakes."

Why did the ancients favor air (or gas, as we would call it today) as the active element in earthquakes? For one thing, they posited a close connection between seismic and volcanic phenomena. They believed that volcanic eruptions, which clearly involve gas, provided an outlet for the forces that would otherwise generate earthquakes. The presence of subterranean air was generally accepted in antiquity. Seneca, for example, had "no doubt that a great quantity of air lies within the underground."

Despite much confusion about the action of underground gases, there appears to have been a strong observational basis for the theory. The ancients could not, of course, have known that the "air" emanating from below entailed some inert gases that we today call carbon dioxide, nitrogen, and helium, but they would have been well aware of the physical manifestations of eruptions of invisible vapors, such as sediments flung upward from the ground. The ancients could not have distinguished the several flammable gases that we now classify as methane, hydrogen sulfide, and hydrogen—some of which, when seeping out at rates and under conditions not susceptible to combustion, would also have caused visible hazes in the atmosphere, or fogs along the ground, as well as asphyxiations of animals. But the ancients were well aware of these physical manifestations of eruptions of invisible vapors, and they looked on, in horror, as flammable gases ignited (being lit, we now know, by electrostatic sparks caused by the friction of fast-moving grains against the rock).

The accumulated observations maintained in folklore and contemplated by the intelligentsia of the time meant that the ancients recognized a variety of phenomena that seemed to serve well as *warnings* of an impending quake. In some ways, folklore is of more practical value to residents of earthquake-prone regions than is our modern science. Pausanias, in his description of Achaia, noted that "in winter the temperature in the region in which the earthquake will occur will suddenly rise. In summer there is a tendency to form a haze, and the sun presents an unusual color. . . . Springs of water generally dry up; great flames dart across the sky. . . . Furthermore, there is a violent rumbling of winds beneath the earth."

The rise in temperature in winter, as a precursor of an earthquake, has been noted in many historical reports. The outgassing theory explains the rise in this way: All the pore-space gases from shallow levels in the ground are the first to be expelled at the surface; having a temperature near the annual mean, as in the ground at shallow levels, they will generally be warmer than the surface temperature in winter. In summer earthquake gas will still be detectable, because it brings up some microscopic particles as well as an enriched proportion of carbon dioxide. Carbon dioxide and other heavy gases will tend to hug the ground and seep into valleys, producing an unusual fog. (The wintertime fog would be dominated by the condensation of water droplets, because upwelling gases are saturated with water vapor by the time they enter colder surface air.) These phenomena have been recorded throughout historical times and, as we shall see, were instrumental in the successful evacuation of a Chinese city just before a disastrous quake.

"This is earthquake weather" was the remark a local guide in eastern Turkey made to a startled American seismologist visiting this earthquake-prone region. A strong quake followed within hours. Evidently this local guide had seen such earthquake fog before, and his ability to predict an imminent quake was better than that of the sophisticated instruments deployed for just this purpose.

In Roman times many writers expressed great interest in earthquakes and collected reports about the attendant phenomena. Pliny discusses precursory effects in his *Natural History,* noting that one sign of an impending earthquake is that "water in wells is muddier and has a somewhat foul smell." He goes on to speculate that caves afforded "an outlet for the confined breath" and that where caves were not a natural endowment of the landscape, constructed tunnels were helpful in dissipating the upwelling air. "Buildings pierced by frequent conduits for drainage are less shaken," he concluded, as are those "erected over vaults."

Seneca noted that before an earthquake, "a roaring noise is usually heard from winds that are creating a disturbance underground." He went on to observe that "often when an earthquake occurs, if only some part of the earth is broken open, a wind blows from there for several days, as happened—according to reports—in the earthquake which Chalcis suffered." Seneca was moved to write his work on earthquakes by a seismic shock that wrecked Pompeii sixteen years before the even greater disaster of the eruption of Vesuvius. I reproduce here one peculiar detail Seneca offered, which today we would explain as having been caused by an upwelling stream of vapors rich in carbon dioxide.

> I have said that a flock of hundreds of sheep was killed in the Pompeian district. . . . The very atmosphere there, which is stagnant, . . . is harmful to those breathing it. Or, when it has been tainted by the poison of the internal fires and is sent out from its long stay, it stains and pollutes this pure, clear atmosphere and offers new types of disease to those who breathe the unfamiliar air. . . . I am not surprised that sheep have been infected—sheep which have a delicate constitution—the closer they carried their heads to the ground, since they received the afflatus of the

tainted air near the ground itself. If the air had come out in greater quantity it would have harmed people too; but the abundance of pure air extinguished it before it rose high enough to be breathed by people.

Seneca sought an explanation for the series of aftershocks that occurred in the Pompeii earthquake, which were felt for several days throughout Campania. He concluded that not all the air had been expelled in the initial eruption but rather that some was still wandering around underground, even though the greater part had been emitted.

Isaac Newton also subscribed to the view that earthquakes were connected with gases. He wrote that "sulfurous streams abound in the bowels of the earth and ferment with minerals, and sometimes take fire with a sudden coruscation and explosion, and if pent up in subterraneous caverns, burst the caverns with a great shaking of the earth, as in springing of a mine." Also noteworthy is that the first edition of the *Encyclopaedia Britannica* in 1771 contained this entry: "Earthquake: in natural history, a violent agitation or trembling of some considerable part of the earth, generally attended with a terrible noise like thunder, and sometimes with an eruption of fire, water, wind."

John Michell, a brilliant scientist of the eighteenth century, made a major contribution to the understanding of earthquakes. He identified a type of earthquake disturbance consisting of slow, ocean-like waves that could actually be observed moving along the surface of the ground. These "visible waves" cannot be explained in terms of elastic wave motion, which would be much faster, and there is not much discussion of them in modern seismological texts. Michell attempted to explain the waves in terms of an eruption of vapor, and that may indeed be the best explanation. What would happen if a burst of high-pressure gas from a depth of many kilometers, and therefore with a pressure of thousands of atmospheres, were suddenly released through fissures in the bedrock into a region beneath a relatively impervious layer of soil that is not brittle enough to develop fissures? Michell reasoned as follows:

> Suppose a large cloth, or carpet (spread upon a floor), to be raised at one edge, and then suddenly brought down again to the floor; the air under it, being by this means propelled, will pass along, till it escapes at the

opposite side, raising the cloth in a wave all the way as it goes. In like manner, a large quantity of vapor may be conceived to raise the earth in a wave, as it passes along between the strata, which it may easily separate in a horizontal direction, there being . . . little or no cohesion between one stratum and another.[7]

Evidence for the phenomenon of visible waves in numerous earthquakes in ancient and modern times is indisputable. Where an earthquake is felt both on exposed basement rock and on alluvial fill, the visible waves are reported on the alluvium only. Alluvium is the sorted, and in places very fine, sediments deposited in the flood plain of a large river or in tidal mud flats. The fine-grained nature of the moist sediments easily traps upwelling gases for a time, and the suppleness of the muddy material enables a good deal of displacement to occur without fracturing. In many cases, large displacements of these waves across stretches of alluvium seem to have wrought more destruction than the sharp shocks of the quake. It is likely that the blanket of alluvial sediments is genuinely lifted off the basement rock by upwelling gases, making it subject to flexural gravity waves, just like the carpet in Michell's example.

Michell's attention was directed to earthquakes as a result of the disastrous one that struck Lisbon in 1755, and he drew from a large number of eyewitness accounts that appear to link these earthquakes with gas. He writes about the flames from the earth and the peculiar fog that accompanied the Lisbon earthquake. Michell also describes precursor phenomena in Jamaica and New England that occurred two or three days before earthquake events; the waters of wells became muddy and developed a sulfurous odor.

Moving into the nineteenth century, I will mention a powerful earthquake that struck a section of the United States and for which conventional theory offers no good explanation. That earthquake was actually one of a series of major and many minor quakes that occurred over a period of several months during the winter of 1811–1812. The site was New Madrid, along the west bank of the Mississippi River in the southeastern corner of Missouri. Significantly, from the standpoint of the upwelling theory of coal (and swamp) formation presented in Chapter 5, much of the disturbance there was expressed in the area known as the St. Francis swamps. Surface phenomena that accompanied this quake were reported in detail in the 1858 *Annual Report* of the Smithsonian Institution.

On the 16th day of December, 1811, at two o'clock in the morning, the inhabitants of New Madrid were aroused from their slumbers by a deep rumbling noise like many thunders in the distance, accompanied with a violent vibratory or oscillating movement of the earth from the southwest to the northeast, so violent at times that men, women, and children caught hold of the nearest objects to prevent falling to the ground.

It was dangerous to stay in their dwellings, for fear these dwellings might collapse and bury them in their ruins; it was dangerous to be out in the open air, for large trees would be breaking off their tops by the violence of the shocks, and continually falling to the earth, or the earth itself opening in dark, yawning chasms, or fissures, and belching forth muddy water, large lumps of blue clay, coal, and sand, and when the violence of the shocks were over, moaned and slept, again gathering power for a more violent commotion.

On this day twenty-eight distinct shocks were counted, all coming from the southwest and passing to the northeast, while the fissures would run in an opposite direction, or from the northwest to the southeast.

[The Pemiseo River] ran a southeast course, and probably was either a tributary of the St. Francis or lost itself in those swamps. This river blew up for a distance of nearly fifty miles, the bed entirely destroyed. . . . The earth, in these explosions, would open in fissures from forty to eighty rods [660 to 1320 feet] in length and from three to five feet in width; their depth none knew, as no one had strength of nerve sufficient to fathom them, and the sand and earth would slide in or water run in, and soon partially fill them up.

Large forest trees which stood in the track of these chasms would be split from root to branch, the courses of streams changed, the bottoms of lakes be pushed up from beneath and form dry land, dry land blew up, settled down, and formed lakes of dark, muddy water.

Where the traveled, beaten road ran one day, on the next might be found some large fissure crossing it, half filled with muddy, torpid water.

It was dangerous to travel after dark, for no one knew the changes which an hour might effect in the face of the country, and yet so general was the terror that men, women, and children fled to the highlands to avoid being engulfed in one common grave. One family, in their efforts to reach the highlands by a road they all were well acquainted with, unexpectedly came to the borders of an extensive lake; the land had sunk, and water had flowed over it or gushed up out of the earth and formed a new lake. The opposite shore they felt confident could not be far distant, and they traveled on in tepid water, from twelve to forty inches in depth, of a temperature of 100 degrees, or over blood heat, at times of a warmth to be uncomfortable, for the distance of four or five miles, and reached the highlands in safety.

On the 8th of February, 1812, the day on which the severest shocks took place, the shocks seemed to go in waves, like the waves of the sea, throwing down brick chimneys level with the ground and two brick dwellings in New Madrid, and yet, with all its desolating effects, but one person was thought to have been lost in these commotions. A family of the name of Curran were moving from New Madrid to an old French town on the Arkansas River, called the Port; had passed the St. Francis swamps and found some of their cattle missing; Le Roy, the youngest son, took an Indian pony, rode back to hunt them, and was in the swamp when the first shock took place, was never seen afterwards, and was supposed to have been lost in some of those fearful chasms.

The Smithsonian report of the New Madrid earthquake also describes some interesting phenomena (including precursor phenomena) that are unquestionably associated with gas emanations.

The morning after the first shock, as some men were crossing the Mississippi, they saw a black substance floating on the river, in strips four or five rods [66–82 ft] in breadth by twelve or fourteen rods in length [198–231 ft], resembling soot from some immense chimney, or the cinders from some gigantic stove-pipe. It was so thick that the water could not be seen under it. On the Kentucky side of the river there empties into the Mississippi River two small streams, one called the Obine, the other the Forked Deer. Lieutenant Robinson, a recruiting officer in the United States army, visited that part of Kentucky lying between those two rivers in 1812, and states that he found numberless little mounds thrown up in the earth, and where a stick or a broken limb of a tree lay across these mounds they were all burnt in two pieces, which went to prove to the people that these commotions were caused by some internal action of fire.

About four miles above Paducah, on the Ohio River, on the Illinois side, on a post-oak flat, a large circular basin was formed, more than one hundred feet in diameter, by the sinking of the earth, how deep no one can tell, as the tall stately post-oaks sank below the tops of the tallest trees. The sink filled with water, and continues so to this time.

Even today, nearly two centuries later, some of these mounds and sinkholes can still be seen. Most conspicuous is the feature called "sand blows," which are funnel-shaped depressions in the alluvial ground, where sand below the soils and clays was blown up and out (Figure 8.1). It is worrisome to note that other sand blows can be found in the region, and their present vegetation indicates that these blows developed some three hundred years earlier. The New Madrid earthquake of recorded history was therefore not the first. What might the future hold?

Turning to the twentieth century, we find that many fascinating (and some undoubtedly credible) eyewitness reports of the great San Francisco earthquake of 1906 and numerous other earthquakes have been filed, especially in the popular press, and that these include accounts of the same sorts of gas-related phenomena. Among the most interesting is a report of the earthquake that ravaged the Hai-cheng region of northeastern China in 1975. This story is particularly fascinating because Hai-cheng was successfully evacuated *two hours* before the 7.3 magnitude quake struck. How could it have been predicted?

Liao-ling Province Meteorological Station reported that in the weeks preceding this earthquake, the air temperature in the vicinity of the Hai-cheng fault was higher than in the surrounding region. This difference increased at an accelerating rate up to the day before the quake, when the differential reached a full 10°C. According to the report filed by the meteorological station,

> During the month before the quake, a gas with an extraordinary smell appeared in the areas including Tantung and Liao-yang. This was termed

"earth gas" by the people. . . one person fainted because of this. . . . Many areas were covered with a peculiar fog (termed "earth gas fog" by the people) just prior to the quake. The height of the fog was only two to three meters. It was very dense, of white and black color, non-uniform, stratified, and also had a peculiar smell. It started to appear one to two hours before the quake, and it was so dense that the stars were obscured by it. It dissipated rapidly after the quake. The area where this "earth gas fog" appeared was related to the fault area responsible for the earthquake.[8]

Apparently these qualitative phenomena, combined with the temperature data recorded at the meteorological post, were taken seriously enough to prompt an evacuation before the earthquake struck.

"Earth gas fog" that streams out at a sufficiently high rate and under conditions unfavorable to mixing can kill. In 1986 a gas cloud

Figure 8.1 Sand vents (blows) similar to those at New Madrid have been seen in locations of several other strong earthquakes. This photo is of such a fomation at the site of a strong earthquake in India on June 12, 1897. Photo by R. D. Oldham.

(thought to have been largely carbon dioxide) emerged from Lake Nyos, a volcanic lake in Cameroon, West Africa. Some 1700 people and 3000 cattle died from asphyxiation. Helium isotope data demonstrated that the gas had upwelled from mantle depth.[9] Three years later, testing indicated that the lake was rebuilding its carbon dioxide stores. Presumably, another disaster will strike at some time in the future.

Gas discharges from the earth that occur just prior to an earthquake may be too slight to be sensed by humans but may nevertheless be noticed by animals, either by their sense of smell or when asphyxiating gases fill underground burrows. Strange animal behavior is included in many reports of precursor events. Perhaps the earliest such description pertains to the earthquake that destroyed the Greek cities of Helike and Bura on the southern coast of the Gulf of Corinth in the winter of 374–73 B.C. The Roman writer Aelian (circa A.D. 200), in his book *On the Characteristics of Animals,* tells the following remarkable story:

> For five days before Helike disappeared, all the mice and martens and snakes and centipedes and beetles and every other creature of that kind in the town left in a body by the road that leads to Carynea. And the people of Helike, seeing this happening, were filled with amazement, but were unable to guess the reason. But after the aforesaid creatures had departed, an earthquake occurred in the night; the town collapsed; and an immense wave poured over it, and Helike disappeared.

Aelian's rather quaint description of an organized exodus of the town's vermin is no doubt an exaggeration. He was, after all, writing nearly six centuries after the events described, which was more than enough time for the story to take on the embellishments of folklore. Nonetheless, I believe that this story was not simply dreamed up. Rather, it seems likely that *some* highly unusual disturbance of ground-dwelling creatures made an impression on the people of Helike before the earthquake and tsunami destroyed their city. We now know of hundreds of accounts of animals behaving in a similar fashion prior to earthquakes. Such reports come from sources as remote from one another in space and time as ancient Greece and modern China.

A recent example is an account from an eyewitness to the catastrophic Tangshan (China) earthquake of July 1976. The account's
author and his companions were all intellectuals in a "re-education
program" at a state-owned farm outside Tangshan. The time of the
strange animal behavior was around midnight, some four hours before
the earthquake.

> We were telling stories in the dormitory when out of the large dorm
> opposite ours burst hundreds of rats. Back and forth they swarmed,
> many scrambling five or six feet up the walls until they lost hold. All we
> could do was watch until they finally vanished into the darkness. As we
> pondered this in amazement, the sound of thousands of excited hens
> and roosters reached our ears. There was a poultry farm nearby, but
> nobody had recalled ever hearing the roosters crow at night. None of us
> knew that this queer animal behavior foretold the coming of an
> earthquake.[10]

Though filled with amazement—like the people of Helike twenty-
three centuries before—the Tangshan witness and his companions
were apparently not well versed in folklore. They went to bed, and a
few hours later some of them were killed when their dormitory collapsed. Regionally, more than 200,000 people perished.

Earthquake Spots and Earth Mounds

Some places are distinguished not by violent
earthquakes in recorded history but by an
ever-present low level of quake activity.
These *earthquake spots* present surface features and precursor phenomena that strongly support the upwelling-gas theory of earthquakes.

There is an earthquake spot about 12 kilometers in diameter in
northern Norway, where for a long time visitors could be all but guar-
anteed to experience at least one tremor a day. These tremors were
weak earthquakes, just barely discernible to human senses. But they
could not be explained in the usual way. No fault line was present to
which ground slippage might be attributed. The ground just kept shak-

ing. A very similar story comes from two places in the United States. One earthquake spot active in recent years is on the western tip of Flathead Lake in Montana. The other, which has been active for at least eighty years, is in Enola, Arkansas.

Earthquake spots can in no way be explained in the usual fashion; they are clearly not born of plates shearing against one another. Earthquake spots are distant from active tectonic structures. There are no plunging plates of ocean bottom or slide-slipping continental blocks nearby. Moreover, the quake activity is confined to small areas. These are isolated spots, not expansive regions of ground tremors.

I believe that earthquake spots are best explained—indeed, can only be explained—by the upwelling-gas theory. Upwelling spurts of light hydrocarbons, especially methane, along with associated gases such as carbon dioxide, force their way up from great depths, causing fractures in the rock to open and shut repeatedly, marking the passage of these pressurized fluids. Both empirical and theoretical considerations have compelled me to draw this conclusion.

Another North American earthquake spot is found on the north shore of the St. Lawrence River, within a meteorite impact structure called Charlevoix. A large meteorite struck there some 350 million years ago, creating a circle of about the same size (and age) as the Siljan Ring in Sweden. As with Siljan, the meteor hit a region of ancient granitic rock; Charlevoix is within the geological province known as the Canadian Shield. Some tremors that can be felt occur at Charlevoix every few days, and micro-quakes are registered very frequently. In this case, proximity to fault lines, including the major fault line of the St. Lawrence River, complicates the discussion somewhat, because one might attribute the quakes to rock slippage along the fault (rather than, as I shall suggest, to gas emanations working their way upward through the fault). Nevertheless, the concentration of the seismic activity within the confines of the impact structure is quite evident.

I visited Charlevoix on several occasions beginning in 1988—not because of Charlevoix's frequent and anomalous earthquakes but because this impact site was a virtual twin to that of Sweden's Siljan Ring, which had been the focus of my attention for the previous five

Figure 8.2   Earth mounds at a golf course in Charlevoix, Canada.

years (see Chapter 6). On my first visit, I discovered that the two sites share a most intriguing feature: *earth mounds.* In Charlevoix, clusters of rounded, steep-sided hills rise abruptly out of an alluvial plain. These mounds are from 2 to 15 meters in height and up to 70 meters or so in horizontal dimensions. They are composed internally just of the clay and sand of the local alluvium, and no satisfactory account of their origin has been proposed.[11]

Two years before my visit to Charlevoix, Marshall Held, a research associate in my department at Cornell, had witnessed the same earth feature. In 1986 he visited the earthquake spot near Enola, Arkansas, renting an airplane to survey the area from above. To his amazement he saw (and photographed) a dense cluster of mounds in an otherwise smooth alluvial plain. It would be a strange coincidence if earthquakes and mounds were unrelated effects and yet occurred together within a patch of only a few kilometers in Enola, Charlevoix, and other earthquake spots.

Methane emanating from such mounds supports the theory that, at least in these instances, puffs of upwelling gases are the cause of persistent tremors. The Charlevoix mounds must have formed between the end of the last Ice Age (the glaciers would have scraped away any mound) and development of the golf course. They may, in fact, be representative of larger features—mud volcanoes (discussed earlier)—that are also strongly related to earthquake activity.

The mounds at the earthquake spots in Charlevoix and Enola may have been produced in a similar way to the mud volcanoes, just on a smaller scale. Some of them have visible holes on top, and in some cases there is evidence that the ground has deformed in recent times: Trees growing on the sides of the mounds all lean outward from the vertical axis. In contrast to mud volcanos, the form of these earth mounds suggests that sediments may have been bulged into place from below rather than spewed out at the top. In either case, upwelling gases would be the most likely cause, especially in view of the present gas seepages.

Upwelling Deep Gas as the Cause of Earthquakes

The Western scientific view is that earthquakes are caused by the same kinds of tectonic stresses that are believed to have shuffled massive blocks of continental and oceanic plates in the course of time. This assumption, coupled with the preference for data collected by precise and impersonal seismographs, means that eyewitness accounts like those cited earlier are usually of little interest to Western scientists, and their existence is not even known by many seismologists. In China, Japan, and the Soviet Union, however, much more attention is paid to gas phenomena. Japan even has a "Laboratory of Earthquake Chemistry." The United States is far behind in this field, not because it lacks the technology, but because it took a wrong turn some time ago and is not open to a change in course.

Surely, however, the citizens of earthquake-prone regions will be more concerned with obtaining a timely warning than with taking

sides in a scientific controversy. Observations of the activity of subsurface gases—such as changes in ground-water levels in water wells and
changes in gas composition or pressure above a water table—are simple and comparatively inexpensive to make, and they can be obtained
objectively. To my mind, it is high time that California and the Central
Mississippi region acquire the knowledge and experience in this field
that will make meaningful prediction possible. Instrumentation oper-
ated by scientists should be one aspect of an early-warning system;
public earthquake education and a reporting network should be
another. In tandem, the two would ensure the widest possible coverage
for the observation of the many phenomena—qualitative as well as
quantitative—that may be relevant for predictions.

The public-safety issues are too important for research to be limited to only one of two reasonable explanatory paradigms. I suspect
that over time, it will be shown that *both* views have relevance. The old
theory (which I wish to resuscitate) that earthquakes are caused by the
movement and eruption of gases can be melded with the modern theory of crustal block movements. Together they would give a much better explanation of all the phenomena than either theory can do alone.

Thus earthquakes may best be understood in terms of a combination of strain in the rocks and upwelling fluids.[12] The build-up of strain
in different rocks would occur unevenly in different locations and at
different depths. But the strain theory alone cannot account for all
earthquakes, especially those that occur at great depth. Rocks deeper
than about 60 kilometers flow plastically, rather than breaking suddenly when a critical stress is exceeded. The internal friction opposing
shear flow is greater than any mechanical strength. Yet earthquakes are
known to occur at depths down to 700 kilometers, deep into the mantle. A recent quake (June 8, 1994) of magnitude 8.2 on the Richter scale
was recorded, deriving from a depth of 600 kilometers below Bolivia.
Clearly another process, not simply the shearing of plates, must be
going on down there. Finding the cause of that process may be the key
to understanding all earthquakes.

In my view, only the presence and rapid flow of large quantities of
gas can be responsible for canceling out the internal friction at depth.

At shallower levels, the invasion of gases from below support and open cracks, thus abruptly weakening the brittle rock. As its ultimate strength decreases, the rock reaches the failure point, and an earthquake occurs. The same mass of fluid may continue to generate earthquakes in its upward travels, as layer upon layer of less permeable rock is encountered, compromised, and then broken. It is not therefore an increase in the stress of the region that brings about the earthquake, but rather a sudden *decrease in the strength of the rock.*

It is not surprising, then, that little evidence can be found through the measurement of strain preceding a major quake. Nothing noteworthy will have happened at the time to the stress field other than the inflation of pore spaces. A bulging of the surface upward might be measurable, but bulging is difficult to observe except at the seashore (where it has indeed sometimes been noticed as a precursor), or in recent times with the Global Positioning System. Such an inflation may introduce very little horizontal deformation on or near the surface that would be detected on strain gauges. If the regions that are inflated are large, even tilt observations will not offer much. On the other hand, the direct gas-related phenomena may be plainly in evidence over the entire region under which the gas has distributed itself.

What precisely should we look for? Leakage of gas driven through small pore spaces by an underlying larger mass of gas will be the first evidence that can be observed. At shallow levels, this could be signaled by disturbances of the ground water and consequent changes in the electric currents this ground water carries and also by changes in the composition of the gas as components from deeper levels are brought up. Unusual noises may ensue. There may even be measurable changes of seismic velocities as pore spaces increase in number and size and the rock is thereby made more compressible. And yes, we should keep a watchful eye out for erratic behavior in our companion animals, whose noses are more sensitive than ours, and in the animals that dwell in burrows and tunnels beneath the surface, where the composition of the gases may suddenly change and become unsuitable for supporting animal life (Figure 8.3).

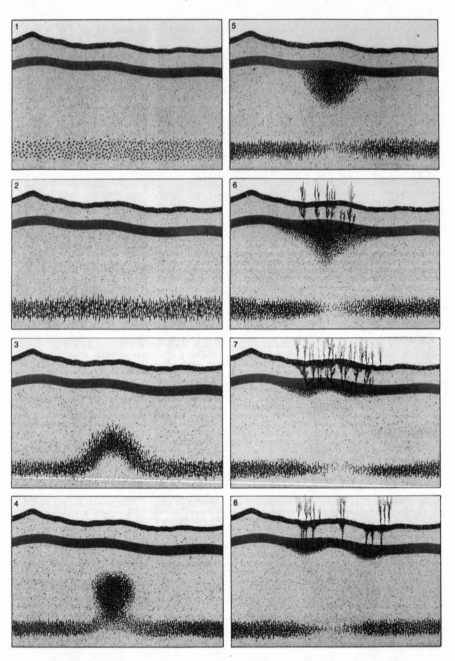

Figure 8.3 Upwelling fluids as the cause of earthquakes. Fluids liberated from the earth's original store of gases, including hydrocarbons at a depth of perhaps 150 kilometers, create pore spaces in the surrounding matrix of hot rock (1). Because the rock at such depths is hot enough to deform plastically, pore spaces distend, often gradually and without sending out the shocks of an earthquake (2). But the instability of the light fluids in the denser rocks is not

Improved methods for earthquake prediction are thus one benefit we may garner from the upwelling-gas theory of earthquake origins. We cannot, however, expect this theory to attract much attention among earthquake specialists until the foundationl deep-earth gas theory is more widely entertained.

As noted in Chapter 7, the deep-earth gas theory and perhaps its companion, the deep hot biosphere theory, may open up new paths for understanding how valuable mineral deposits are formed. Earlier chapters presented the profound changes in petroleum geology that such a new paradigm would demand, along with the profound changes it would necessitate in our understanding of life *within* the earth. The final two chapters will explore more speculative implications of the two theories. How might the deep-earth gas and deep hot biosphere perspectives alter our scientific views of the origin of earth life? And what do these two linked theories mean for our chance of encountering life elsewhere in the solar system?

Figure 8.3 *(continued)*
relieved (3, 4), and the fluids are driven farther upward. At shallower levels where the rock is harder and no longer plastic (dark band in 5), the fluids cause brittle fracture of the rock. Small cracks induced by fluid pressure develop and grow, weakening the rock. As its ultimate strength decreases, the rock eventually reaches the failure point, and this causes the earthquake. It was not any critical rise in the stress of the rock that was the immediate cause of the quake; rather, rapid weakening of the strength of the rock initiated the quake (7). Gas that is present in a wider region around the epicenter, and that did not escape at the time of the event, continues to weaken other rocks until they also give way. This explains the usual widening area of aftershocks. SOURCE: After Thomas Gold and Steven Soter, 1980. "The deep-earth gas hypothesis," *Scientific American* 242: 154–61.

Chapter 9   The Origin of Life

The earth supports not one but two large realms of life: surface life fed by photosynthesis, which is familiar to us all, and deep life, fed by chemical energy that has penetrated up from below. We have only just begun to explore the inhabitants and the reach of the deep realm. I suspect that until microbes drawn up from the deep are perceived as representatives of a wholly distinctive biosphere, rather than as isolated and ingenious adaptations of surface life pushing back the frontiers of habitability, research on deep life will remain sparse and largely unheralded. If the shift in perspective does take place, however, a veritable explosion of new ideas will surely permeate two of the most speculative yet philosophically engaging issues in science: the origin of life and the prospects for extraterrestrial life.

Detailed chemical analysis shows that expressions of life in the surface realm and in the deep realm almost certainly have a common derivation, because both have the same genetic system. Whether the shared genetic system is due to a common panspermia—the transport of biological materials from other astronomical bodies to the earth—or, alternatively, life arose in one of the two realms on this planet and then evolved adaptations that enabled it to populate the other, we do not yet know for certain.

If the origin of earth life was in fact terrestrial, then we would want to determine in which of the two realms it originated. One way to begin this inquiry is to search for dependencies between the two realms. Contact between the surface and the deep is certain to have been present, but contact is very different from dependence. If we can find evidence of such a dependence of one realm on the other, but not of their codependence, we will have a strong case for arguing that life arose in the independent realm and later spread to the dependent realm. But if no distinctive dependence can be found, then we must seek some other considerations that might indicate in which direction evolution had gone.

There is (as yet) no evidence on the nature of that sequence or on the relationship these two realms have had with each other over time. They may be essentially independent of one another at present. If all the photosynthetic surface life were to disappear, for example, the deep subsurface life might continue essentially as before. Similarly, if for some reason deep life were to disappear, we know of no reason why this would have much impact on the photosynthetic surface life—at least in the short term. (It might make a difference in the long term, because there may occasionally be beneficial exchanges of genetic material between the microbial life at depth and the surface life.)

The Habitability of Surface and Subsurface Realms

A fruitful way to approach the question of the origin of life is to compare the habitability of the two realms. The surface life we know is enormously rich because of the large amount of energy that the sun has provided to photosynthetic microbes and, eons later, to photosynthetic algae and land plants. These organisms have evolved sophisticated apparatus for converting solar energy into chemical forms, on which they and the rest of the food chain of the surface biosphere then depend. Photosynthesizers have adapted to the absence of energy during the night and during seasonally lengthened nights at high latitudes. From time to time, surface life must cope with a far more severe problem: blockage of sunlight for periods of months or more in the aftermath of severe volcanic eruptions and occasional impacts of large asteroids. Microbes can wait out these catastrophes in a quiescent state, but global populations of macroflora and fauna may be severely challenged by these events, resulting in large extinctions.

Is the energy supply that supports the deep hot biosphere subject to the same risk of change and blockage? We do not know how steady and long-lasting upwelling streams of hydrocarbons are in any particular location, but I suspect there would be great constancy over millions of years. Any disruption is unlikely to affect the entire subsurface realm of the planet all at once. Moreover, because the deep biosphere (with the exception of life at the borderlands, such as oceanic vents) is entirely microbial, long enforced periods of quiescence may be no great challenge. Then, too, impacts of large asteroids may have devastated the surface, especially before 3.8 billion years ago. But such impacts may actually have benefited the subsurface biosphere by opening up new cracks for hydrocarbons to upwell, allowing microbial life to flourish.

How do the two realms compare in temperature suitability for life? As explained in Chapter 2, temperatures that would boil water at the earth's surface are easily tolerated by microbes at depth, because high pressure substantially raises the boiling point of water. Liquid water is thus available at depth over a wider range of temperatures than it is on the surface of the earth. With the exception of areas of active volcanism, the temperature gradient of the subsurface earth holds steady at any given depth; there is no "weather" below the surface and no glacial episodes or boiling water. This is in sharp contrast to conditions on the surface—particularly on land—where enormous temperature shifts may occur seasonally and even daily. For the surface to have remained habitable over the long course of geological time, temperatures must have remained within a narrow band conducive to water remaining liquid at the pressure offered by the earth's atmosphere.

A great advantage for life at depth is the protection the rocks offer from the harsh ionizing radiations both from the sun and from space. We surface creatures tend to forget that surface life has evolved protective coatings and pigments to ward off harmful solar radiations, as well as DNA-repair mechanisms to undo the damage inflicted by solar and cosmic radiations that cannot be completely avoided. But none of these adaptations would have been available to the very first life forms—to the very first attempts at cellular life. In contrast, subsurface conditions offer a much more agreeable situation. There is no need at depth for complex molecules to defend themselves against the radiations that bombard the surface. It is true that hard ionizing radiations would also exist below as a result of the radioactive decay of potassium, uranium, and thorium. These radiations, however, would suffuse the rocks at a much lower rate than that experienced on the surface, and the radiations would hold steady over long periods.

Finally, the most important difference between the deep hot biosphere and the surface biosphere as realms for the origin of life is the abundance of primordial energy upwelling from below—and, just as important, from a depth so distant that the source itself is inaccessible to life. Remember that the first living cell could not have performed the feat of photosynthesis. On the surface, a large chemical energy supply would thus commence only after some organisms had developed the complex actions of photosynthesis. Before then, the spontaneous assembly of molecules that can undergo energy-yielding reactions would have been a rare event, and concentration of such molecules into a "primordial soup," perhaps sloshing around in a tide pool, would have been even rarer. Should a self-maintaining and self-reproducing system of complex molecules have arisen in such circumstances, what would have happened to this pioneering form of "life" when all the food was used up at the exponential rates at which living systems reproduce? It is not clear how this "feast and famine" situation could have been avoided before the advent of photosynthesis or how a long and detailed evolutionary process could have been sustained.

For this reason alone, the subsurface is the more likely location for the early phases of life if the deep-earth gas theory is valid. Hydrocarbon fluids streaming up from below, and originating at depths far too hot for carbon-based life to reach and plunder, might have offered sustenance at a steady, metered rate for long periods. This scenario would provide ideal conditions for life to arise and flourish. Later, the same conditions would have allowed microbial life to develop a range of chemical abilities. Motile adventurers at the outer edge of the subsurface realm—perhaps in the vicinity of vents on the deep ocean floor— might have developed heat-sensing pigments by which to orient themselves toward the energy-rich vent, thus preventing their being carried away into the barren, cold ocean. It has been suggested that the first important step toward photosynthesis was orientation and navigation by the sensing of heat radiation.[1] Photosynthesis would have followed if some microbes found it advantageous to live at or near the surface and to enrich their energy sources by the use of sunlight.

Going in the other direction, from surface to deep life, I cannot see a similarly favorable situation. Photosynthesis would have to develop, as a very early step, from an evolution dependent on some unspecified form of chemical energy becoming available continuously on the surface. The progression would then have to take the form of an invasion of the subsurface, the development of an ability to use the chemical energy sources available there, and an accommodation to the elevated temperatures and extreme pressures of that realm.

The subsurface realm, for a number of theoretical reasons, therefore appears to be the more likely site for the initial development of this curious and extremely elaborate chemical processing that we call life. Pressure and temperature conditions in the subsurface are steadier and generally more conducive to life than are surface conditions. Subsurface life would not only have tolerated large asteroid impacts but would probably have benefited from the disruption. Radiations harmful to life would have been greatly reduced at depth. Finally, according to deep-gas theory, chemical energy would have been abundant and supplied as a metered flow.

Empirical support for the subsurface realm as the locus for life's origin takes the form of many recent taxonomic analyses that identify hyperthermophilic archaea and bacteria as the most "deeply rooted"— the most ancient—life forms. It is of course possible (though very unlikely in my view) that the earth's surface some three billion years ago was fit only for the most highly heat-loving forms of life. The deep-rootedness of extreme thermophiles might also be taken to suggest the hot ocean vents as the locus of life's origin. But what would have been the source of chemical energy, if not fluids upwelling from the deep that could derive energy in reactions with materials available on their pathways? Empirical evidence, then, though it does not exclude hypotheses that posit a surface origin of life, nevertheless strongly supports the contention of subsurface origins.

The Enhanced Probability for Life's Origin

Deep-earth gas theory is a requisite for the view that earth life originated at depth. The subsurface realm not only provides a more favorable habitat for early life but also vastly increases the region in which chemical "experiments" could have taken place. This increase in volume and mass—especially of carbon-carrying molecules, greatly enhances the probability that favorable co-occurrences of chemical reactions and juxtapositions of complex molecules—the precursors of life—would have happened by chance.

It may be quite incorrect to think that there was a definite beginning to life. There may be a step-by-step route toward the complexity of very simple forms of life in which no single step could be considered *the* step that demarcates living from non-living matter. Life may represent no more than the processes that are described in the physics and chemistry textbooks, applied in circumstances that are far outside the scope of our imagination. We must therefore judge the probability or improbability of any particular molecule or structure forming in terms of the number of experiments that could be expected to have taken place in the chemical media characteristic of the early earth—and in time spans of hundreds of millions of years. A process that would rightly be judged highly improbable for the occurrence of any single step would nevertheless have a high probability of taking place if the experiment that could produce it were repeated a very large number of times.

Suppose we start with the variety of atoms (or molecules derived from them) that we see in liquids in the deeper rocks: carbon, hydro-

gen, oxygen, nitrogen, sulfur, phosphorus, sodium, potassium, calcium, and some other metals derived from the rocks. (I will refer to this
assortment as the "soup.") How many reactions would be taking place
that might create novel molecules? The subsurface realm offers two
fundamental advantages over the surface realm in this regard. First,
pressure would stabilize many molecules that could not exist at the
surface. More novelty would thus be possible at depth, given the same
starting brew. Second, the elevated temperatures associated with depth
would speed all reactions, thus offering more possible combinations
than a similar volume of soup in the surface realm would generate over
the same span of time.

How might the volume of soup differ between the surface and subsurface realms? Proponents of the surface view of life's origin do not
hold that the entire surface of the earth was awash in life-generating
soup some 3.5 or 4 billion years ago. Rather, in their view, molecules
that could serve as precursors to living metabolisms and replication
systems might have been manufactured sporadically by chance and
then concentrated in sloshing and evaporating tide pools or along
vents of the deep ocean floor, where hot fluids spew into a cool and
chemically distinctive environment. According to deep-earth gas theory, however, upstreaming hydrocarbons would have been suffusing
pore spaces *within* the earth's crust with a ready-made soup ever since
the planet accreted and gravitational sorting commenced. The volume
of soup available for prebiological experiments would thus have been
virtually the entire volume of pore spaces held open within the earth's
crust by the upwelling fluids.

To calculate the possible volume, let us begin with a reasonable
interval of depth, say from just below the surface down to 10 kilometers, in which the chemical soup might fill pore spaces that occupy 1
percent of the rock volume. The rock volume would amount to 5.1×10^{18} cubic meters; 1 percent would thus be 5.1×10^{16} cubic meters, or
a mass of soup on the order of 5.1×10^{16} tons. This mass of soup would
represent about 4 percent of the entire mass of the earth's oceans today.

Let us assume that the mean molecular weight of the soup is 50
atomic mass units. This would entail, say, two carbon atoms (24), one

oxygen atom (16), and ten hydrogen atoms (10). Fifty atomic mass units is not too different from the molecular weight of liquids pulled up from such depths today. Then, 5.1×10^{16} tons would translate into a number of such molecules on the order of 6×10^{44}.

How many chemical reactions would take place in the soup? Here the calculation becomes very speculative. We can infer (as explained in Chapter 3) that because the soup comes from an underlying source in the mantle that has not been heated and mixed to the point of chemical equilibrium, it will possess energy that can naturally drive many chemical processes that take it nearer to chemical equilibrium. Let us suppose that any one molecule has a chance of suffering a modification once per day. Such modification might be just thermally induced, or it might result from chance encounters of molecules that will react with one another. In a billion years there will then have been 3.6×10^{11} modifications for each ancestral molecule present at the beginning, in a total mass of 6×10^{44} such molecules, for a total of about 2×10^{56} such modifications.

This large number would assume importance only if some special molecule crucial to life were created among the crowd of others. How can we evaluate the possibility that this might happen? Probabilities so small that the particular molecule would occur only once in 10^{56} experiments are not in the realm in which we have any competent intuitive judgment. Perhaps an analogy may help, however. Imagine that you are flipping a coin in sets of 100 tosses. After you flip the first hundred, you decide to flip another hundred, and then another, until you finally flip a set in which all one hundred tosses turn up heads. If you have average luck, you would have to perform that set of tosses 10^{30} times in order to achieve a single all-head set. Now, if you were given 10^{56} opportunities to perform the same experiment in coin-tossing, you could expect to produce, on average, 8×10^{22} perfect sets of all heads.

This could also be phrased in terms of the traditional discussion of high improbabilities—that is, of how many monkeys with typewriters it would take to duplicate one of Shakespeare's works. As it turns out, our number here, 10^{56}, is about the number of monkeys needed to produce just the first line of a particular Shakespeare sonnet.[2] But if we employed 10^{57} monkeys instead, we could expect several instances of success. The absurdly improbable event would have become highly probable.

Whether the numbers I have used in my calculation of molecular probabilities are actually correct I do not know. But whatever the correct choice of numbers may be, it is certain that they would still lead us into this inscrutable large-number realm—and thereby to a situation where events will occur that individually would have been judged so extremely improbable that no notice would be taken of their chance of occurring. But they may indeed occur, given this background of steady modifications in masses upon masses of molecules over a very long stretch of time. In fact, they may have high probabilities of occurring.

Even so, one may wonder what good it would do for the magic molecule to have lurched into existence just once in a billion years amid such a mass of useless companions. The answer is that a single molecule can indeed come to dominate the population of the entire soup in a short time—if that special molecule is an autocatalyst. An autocatalyst is a molecule that not only is responsible for catalyzing a particular reaction but one that causes another molecule just like itself to form from components of the soup. An ordinary catalytic molecule facilitates the formation of some other molecule. An autocatalyst stimulates the formation of a copy of itself.

Let us assume, for convenience, that in all cases the average generation time for a single molecule to form a new molecule is one day. After one day, the ordinary catalytic molecule will have synthesized one molecule of another kind. After two days it will have synthesized one more, for a running total of two synthesized molecules. By the end of the third day, there will be three such synthesized molecules, four on day four, and so on. In contrast, the population of autocatalysts will grow exponentially, from two at the end of the first day to four at the end of the second, then to 16, 32, and so on. This rapid growth in numbers inherent in the autocatalytic process makes its furtherance invulnerable to accidents. The loss of any single autocatalytic molecule among a large set of such molecules is no great setback. In contrast, if the lone catalyst meets an untimely end on the fifth day, the entire process comes to a halt.

The number of offspring of the autocatalytic molecule will be given by 2^n, where n is the number of generations that have passed. Let us take 50 generations—in our model, 50 days—in which period the ordinary catalyst would have produced 50 molecules. In this same period the autocatalyst would have produced 2^{50}, or 10^{15}, offspring. In 188 generations (in our example, still only 188 days), the number would reach 4×10^{56}—which, remember, we have calculated to be the total number of molecules in the entire soup. In practice, there will be shortages of some component atoms of the autocatalytic molecule well before the entire soup is reconfigured into clones of that busy molecule, so the process will stop far short of total consumption of the soup. Nevertheless, this autocatalytic process will easily dominate all other chemical processes, having started with just a single molecule, whose probability of forming might have been considered extremely low.

The more extensive the volume of the soup, the higher the chances of the creation of a molecule that fulfills the complex requirements of autocatalysis and the higher the chances of a further development from that stage. The layers of the crust, in combination with the flow of fluids that can deliver energy by undergoing reactions with the solids on their pathways, are the largest such domain that the earth possesses. We should then give this domain our first consideration for the origin of life.

We can identify autocatalytic processes readily if a molecule produces one like itself in the next generation. But a process still belongs to the mathematical regime of autocatalysis if it reproduces the first type only from a later generation, having in the meantime produced one or several intermediate stages. These intermediate stages slow up the process, but the underlying exponential will eventually still beat out all competing processes. We may refer to autocatalysts of different orders: The first order just reproduces itself. The second order produces itself again, but only after another form intervenes. A third order has two forms in between, and so on.

But does this account not describe the essentials of life? Each form will produce an autocatalyst again, which we will call the *genotype,* the unit that contains the instructions for the design of the next generation. In between is another stage—the *phenotype*—which will then produce the genotype like the one from which it derived. The next genotype then continues the process. Plants and animals (such as ourselves) are the intervening phenotypes. But the mathematics of reproduction will still follow the exponential law. With the power of this exponential law behind it, any biological regime would readily become dominant over any competing one that did not enjoy this mathematical advantage. All reproduction in biology is subject to the exponential law, even when unfavorable for a species, as it is when it leads to the "feast and famine" disaster.

An autocatalyst of the second or higher order presents an opportunity for the "phenotype" stage to support changes that would not destroy the autocatalytic ability of the next genotype but would nevertheless carry forward the changes that had been made. At this point Darwinian logic enters the system, and changes that are "beneficial" by increasing the survival chance of either genotype or phenotype or by increasing its reproduction rate will beat out other changes.

Has this discussion solved the problem of the origin of life? I do not think so. What it has told us is that the basic systems of multiply repeated self-reproduction may be present already in molecules that occur in what we would call inanimate nature; perhaps they even occur frequently. We have thus set the scene for an evolution that will now be subject to the guiding force of Darwinian natural selection and, so enhanced, may expand into much greater complexities both of structure and of chemistry.

The self-replicating molecules are a step toward the evolution of life, but our definition of life generally involves the presence of cells. Perhaps this is an error on our part. Viruses are units of life, but only those that use cellular organisms as hosts and thereby cause changes or damage to them are identified. There may be a large number of similar non-cellular organisms that draw on chemical energy sources from their environment that may be either of biological or non-biological origin. If they do not cause any evident changes, they would give us no hint of their existence. Yet they might have represented the early forms of life and may still be a major component.

Cells appear to be a necessity for the more complex, more elaborate life forms. How could cells ever have formed? Many investigators have regarded the step to the formation of cells as the critical step, the essential step to enter into the evolution of complex forms of life. For this reason, I shall offer a possible path toward the formation of cell walls surrounding the genetic (autocatalytic) molecule. Many microbes are known that acquire water around them, confined and held to them by a jellyfying agent. The non-cellular life may already have developed such an ability, and the next step may be the evaporation of water from the outer surfaces of this slime, thereby concentrating material in the slime, causing them to form a skin. If this proves helpful to the organism it may then evolve and add materials to the surrounding slime that are best suited to make a permanent skin. Once there are molecules of great complexity involved, it is not a large step for these to enclose themselves with a cell wall.

Darwin's Dilemma

In Charles Darwin's lifetime and ever since, one problem has plagued his beautiful theory. The gradual evolution implied by occasional random errors in the genetic blueprints should leave evidence for many intermediate forms in the evolution of a species, some that are in the line of evolution that occurred and some that came to a dead end. But that is not what the fossil record shows. Rather, it shows long periods with little or no change, and then an almost instantaneous transition to another form. It has been argued that this pattern of biological evolution may be due to rare but rapid changes in the environment (possibly imposed by major variations in climate or atmospheric composition or by volcanic eruptions) that enforced a rapid adaptation of life at such times. In between these changes, there was little to be gained from random changes, and Darwinian selection would have tended to inhibit them. This view of biological evolution is generally called "punctuated equilibrium."

Perhaps Darwin's dilemma is in fact resolved by an alternation of stasis and change in the environment. I personally do not think so, although I believe that the environment may well have been subject to such punctuated conditions. But even if there were periods in which a rapid adaptation was needed, could random changes and the numerous selection processes required have taken so little time and left so few examples of intermediate steps? Would this happen to many different species at different times? That, at any rate, is the nature of the question that remains.

Darwin was not very definite about the cause of the variations that his theory required. Most investigators following him assumed that it must be random variations, inaccuracies in the blueprints delivered to produce the next generation. They could certainly identify such errors in many cases, and they most often found these errors by noting their detrimental result. But might there be some other cause for genetic modification than random errors, a cause that could make major changes all at once? If so, the problem of the lack of intermediate forms could perhaps be explained.

Random chance mutation seems to be an adequate mechanism for explaining evolutionary change within two of the three taxonomic domains of life—the archaea and the bacteria. These organisms have high reproduction rates, and the numbers in any generation are enormously large. The probability of hitting on a favorable mutation by random errors in a particular period of time is given by the number of living representatives of a species, divided by the length of time required by the reproduction cycle. In this, the difference between elephants and bacteria is enormous.

Let us consider the evolutionary capabilities of the most-studied bacterium on earth, *E. coli*. There are about 10^{12} *E. coli* bacteria in every person's digestive tract; these microbes are essential for the chemical processing of our food. The human population, rounded to the nearest order of magnitude, consists of 10^{10} persons alive today, so the number of *E. coli* carried by humanity alone is about 10^{22}. A bacterium suitably fed will reproduce about every 20 minutes. Now compare *E. coli*'s reproductive potential with that of elephants. There are perhaps 100,000 elephants alive today, and their reproduction time is 10 years. The bacteria reproduce 262,800 times as fast, and there are 10^{17} times as many in just the one habitat (the human gut) that we have mentioned. The chance of hitting on a favorable mutation in a given period of time is thus 2.6×10^{22} times greater for these bacteria than for the elephants.

Thus chance mutations and their selection could well be the path to major evolutionary innovations among the various lineages of microorganisms and yet be hopelessly slow for large creatures. As it turns out, large creatures differ greatly in form, but not so much in function. A mouse, to say nothing of a frog, may look very different from an elephant, but all vertebrates share the same cell types, the same species of molecules. The innovations among macrofauna occur not in the design of the standard animal set of cell types but in the arrangement of cells and in the speed and number of reproduction cycles that are sustained. Microbes excel in metabolic variety, whereas eukaryotes are the pioneers of form.

Metabolic innovation may have been the more difficult to achieve. A methane-eating, heat-loving microbe may look very much like a photosynthetic microbe, but the evolutionary differences are far greater than those that separate jellyfish and human. Lynn Margulis and others have offered convincing empirical evidence and theoretical arguments that major innovations in metabolism have been the achievements, almost entirely, of the microbial domains. The eukaryotic domain, which eventually spun off all the macrofauna, has simply acquired what it needed from the microbes by way of endosymbiosis—that is, by engulfing microbes that then became integral to the eukaryotic cell. Indeed, the entire eukaryotic domain is thought to have begun with the symbiotic merger of at least two metabolically distinct lineages of microbes.

Endosymbiosis is now accepted as a crucially important and radical means by which evolutionary innovations are passed *across* lineages. Such exchanges can account for sudden leaps in the recipient lineage, even though the microbial lineage that initially pioneered the metabolic talent may have taken a good deal of time to bring it to fruition. Another pathway for sharing skills is now a major enterprise in our biomedical and agricultural laboratories. This is the art of gene splicing. Could naturally occurring methods of gene splicing have been a major driving force for evolution, and could they have resulted in much more rapid transitions and fewer intermediate steps than individual chance mutations would have required? This possibility has recently come under serious investigation by experts in the field.[3] Nevertheless, a little idle speculation from a non-expert may be interesting.

Darwinian logic would apply not only to the evolution of metabolisms and body forms but also to the evolution of the genetic systems underlying both. Thus if any transfer of genes from one organism to another were generally more a help than a hindrance, the genetic system—pioneered by populous and prolific microbes—would surely have adjusted itself to permit this. Even a minute probability of transferring some genetic material from *microbiological* lineages to *macrobiological* ones could be a major—even a dominant—mechanism for *macrobiological* evolution.

To illustrate this point, let us return to our comparison of elephants and bacteria. An enzyme, perhaps a molecule of great complexity capable of performing a useful function in certain metabolic processes, might be "invented" by a lineage of microbes as a result of a long sequence of chance mutations. If these successive chance steps would perhaps take a hundred years to occur somewhere in the large population of microbes, this would translate into 10^{18} years—one billion billion years—for the same chemical evolution to occur in elephants (and similar times for any macrobiological form). In other words, it would never happen. We will now have to think whether there are mechanisms for gene splicing in nature that might seem outrageously improbable but that might still bring a molecule from the microbes into the genetic material of macroorganisms in an acceptable time frame for evolution.

It would seem, then, that all elaborate "design" by a sequence of chance mutations belongs to the two taxonomic domains of microbiology. Mutations in macrobiology may still be a significant factor for simple design changes (including changes in form), though detrimental random mutations occur more often. Chemical changes may well require carrying several intermediate forms that serve no purpose before the useful form is assembled, and that would greatly decrease the probability and therefore greatly lengthen the time required. In my view, only the microbial world can offer up the numbers of experiments that such significant kinds of innovations demand. Gene splicing would then offer the best path for the macrobiological world to benefit, too.

In modern biotechnology we see the two methods at work. Deliberate selection, such as that which made a Pekinese or a Great Dane from a wolf, or a corn of two or three times the yield of its ancestors, is a method still very much in use for agriculture. Strains of plants that resist certain insects or diseases have been produced in this way; dogs, horses, cows, pigs, and other animals have been successfully bred for characteristics desirable to their owners. But now we can already begin to see a competition arising with the other method of modification: the method based on gene splicing. We can, for example, buy gene-spliced tomatoes that exhibit increased resistance to rotting. Gene splicing enables pharmaceutical companies to harvest human insulin from pigs. Many other such advantages are expected to emerge from ongoing experimentation. Agriculture will be able to remain efficient without the chemical assistance it now requires to ward off insect attack and disease, because crops and livestock will themselves have been equipped with genes that endow them with the ability to resist these ravages. How long will it be before we will incorporate disease-resistant genes into humans and diminish our reliance on medications and medical treatments?

If we believe that genetic material could be transferred naturally from one species to another, then the whole subject of symbiosis—the relationship of a set of different creatures living together for their mutual benefit—assumes new importance. Could such colonies have come to pool their genetic material? If so, then by that one act they would become a new, complex, differentiated, multicellular organism. I do not know how far one should take this line of thought. Is a complex animal the descendant of a symbiotic arrangement of the progenitors of the individual organs?[4]

Symbiosis, endosymbiosis, and gene splicing are all ways to spread evolutionary innovations horizontally across great branches of the full diversity of life. For these sorts of exchanges to occur, however, a fundamental compatibility in genetic structure must be present. This brings me to the chapter's final topic: molecular chirality.

Many of the molecules in biological material could exist in two configurations that are precise mirror images of each other. These molecules are said to possess chirality. A chiral molecule may appear in either a right-handed or a left-handed form. The double helix of DNA, for example, resembles a spiral staircase with handrails on each side. In theory, the staircase could spiral either to the right or to the left, and that directionality would hold from whatever aspect one viewed the staircase. In practice, all DNA molecules exhibit right-handed chirality. Why this uniformity throughout all domains of earth life?

A right-hand–left-hand symmetrical relation cannot be defined in two dimensions, so chirality can exist only in molecules of three dimensions. For any three points, we can always choose a plane that goes through them, so a molecule composed of just three atoms will have only two dimensions. A three-atom molecule will always be identical to its mirror image, seen from one side or the other. It cannot possess chirality. Molecules that possess four or more atoms may or may not be three-dimensional and therefore may or may not be chiral. Chiral molecules in a liquid can be observed to rotate, in one direction or the other, the plane of plane-polarized light that is passed through it. Thus the molecules may rotate the plane in a right-handed or left-handed sense. Those that do the former are given the prefix R; those that do the latter, L.

There are many chiral molecules or crystal structures of non-biological origin that occur naturally. But in non-biological materials, the two forms appear in statistically equal numbers. In biological materials that is not so. As already noted, all DNA spirals to the right. Some of the amino acids that form the basic components of proteins are chiral, too, and these show a left-handed chirality for all known terrestrial life—from bacteria all the way to elephants and ourselves. In theory, a right-handed form of any particular protein would undertake precisely the same chemical reactions in all cases in which all interacting molecules had been replaced by their mirror images. In other words, chemical reactions do not favor one chirality over another. Why, then, have not some branches of life shifted their proteins to the R structure?

An obvious answer is that all biology must have derived from the same origin. That first cell—or whatever it was—laid down the rules of the game for all time. The choice of chiral direction was arbitrary, whichever form this progenitor happened to hit on, with a 50–50 chance. Everything that derived from it, however far down the line of evolution, then continued to be constrained by that first selection.

Linus Pauling, one of the great chemists of the twentieth century, expressed his doubts about this answer:

> All the proteins that have been investigated, obtained from animals and from plants, from higher organisms and from very simple organisms— bacteria, molds, even viruses—are found to have been made of L-amino acids. The suggestion has been made that the first living organism happened by chance to make use of a few molecules of the L configuration, which were present with the others statistically in equal numbers; and that all succeeding forms of life that have evolved have continued to use L-amino acids through inheritance of the character from the original form of life. Perhaps a better explanation can be found—but I do not know what it is.[5]

No one can surmise Pauling's thoughts at the time he wrote this, but I imagine he could not believe that inheritance would be so precise throughout evolution that the vast diversity of life would never develop independent branch systems deviating from this defined pattern of chirality. At any rate, that would have been my reservation, leading also to the hope that a better explanation can be found. Perhaps the advantage of gene splicing provides the better answer.

If gene splicing has been a major source of the variances introduced into the evolutionary process, then any lineage that mutated to the opposite chirality would forego all future opportunities to receive the benefits of innovations achieved by any other branch of life. The straying lineage would be cut off from assistance and would, sooner or later, lag behind the rest to such an extent that selection would drive it to extinction.

If gene splicing and various forms of symbiotic mergers across widely separated lineages have, in fact, played major roles in the evolution of earth life, then we should not describe the vast diversity of life through time as an evolutionary "tree," with each branch progressing on its own and developing into individual species. Rather, we would think of a combined evolution of terrestrial biology, all continuing to be closely interrelated with one another and with the most prolific gene pool of all—that of the microorganisms.[6] Because the deep hot biosphere is, in my view, so vast, and because this realm is very likely to have nurtured the first living systems, many of the innovations and gene-trading and merging events that support today's expressions of life probably took place well before there was any life *on* the surface of the earth. Perhaps such achievements are still under way.

Chapter 10 What Next?

A primordial origin of terrestrial hydrocarbons and a source of them at great depth, providing food for a vigorous microbiology at shallower depths—this is the viewpoint for which much evidence has been presented here. It is clear, however, that additional confirmation of the deep-earth gas theory and the deep hot biosphere theory will be required before they become generally accepted. Several types of investigations can be undertaken to acquire such confirming evidence.

The deep-earth gas theory has already been confirmed in the drilling experiment described in Chapter 6. The Siljan Ring geological structure in Sweden was chosen as the site for this experiment because it is a purely granitic province, so any gas discovered at depth there could not be explained by the biogenic theory of the formation of hydrocarbons. Methane and heavier hydrocarbons were, in fact, discovered—and at a depth of more than 6 kilometers. The deep-earth gas theory was thus, to my mind, confirmed. Nevertheless, more such anomalous findings will have to accrue before the existing biogenic theory is abandoned by those who now accept that theory unquestioningly.

Another way to refute the biogenic theory would entail measurements that can be made in existing gas fields. Many observations have

been recorded of hydrocarbons at the earth's surface. "Cold petroleum seeps" on continental shelves and blocks of methane hydrate
ice pressing up through the oceanic floor and underlying vast
stretches of arctic tundra were described in Chapter 2. The La Brea
Tar Pits of southern California, famous for the brown-stained bones of
sabertooth tigers and other great mammals of the Pleistocene, is
another example. In places where hydrocarbons seep from the crust
into the atmosphere in gaseous form and at a high rate, the emanations cannot be seen and may not carry an odor, but flames may
appear and disappear.

Accurate measurements of the rates of gas seepages, particularly
over regions where natural gas is produced commercially, might yield
data that would be difficult to explain by the biogenic theory of hydrocarbon formation. If the volume and seepage rate of hydrocarbon gases
entering the atmosphere in such regions turned out to be so great that
the gas reservoirs underneath would have been exhausted naturally in
just a few thousand years, for example, the conventional theory of
multi-million-year-old gas fields would have to be abandoned. It would
have to be acknowledged that, say, methane molecules now contained
in a reservoir of perhaps late-Cretaceous age have not, in fact, been sitting there, just so, for some seventy or eighty million years. The hydrocarbon content of that reservoir would have to be attributed to a temporary pooling of an ongoing upward flow from a very distant and
much deeper source. We could thus conclude that the resource would
be enormously more productive over time than previously thought.
One more stone would be removed from the foundation supporting the
theory that hydrocarbons are the reworked remains of organisms that
lived and died on the earth's surface, then were buried along with sediments of that particular age, and finally were cooked into hydrocarbons and concentrated into much smaller volumes by mysterious geological processes.

Calculations made using measured values of the permeability of
rocks to the flow of gases seem to show that any gas field would indeed
exhaust itself in a small fraction of the age that is ascribed to it on the
basis of the age of its containment rocks. For example, Jon S. Nelson

and E. C. Simmons and others have calculated that the volume and flow rate of gas seepages that would have occurred from a gas field below rocks of the lowest permeability recorded would still exhaust the field in just a few tens of thousands of years.[1] Direct observational evidence of escape rates will have to be obtained, however, before these seemingly anomalous results are taken as a serious challenge to the biogenic theory.

There is one very inexpensive way to measure the rate of out-gassing in many places. Over a small area, erect a tent filled with a permeable substance like sand and made of an impermeable material like plastic sheeting. Arrange for the outside edge all around to be dug into the ground to a sufficient depth that the wind wouldn't blow through. Then wire the interior of this tent with instruments that continuously measure the composition of the air and other gases. Methane and carbon dioxide would almost certainly be the principal gases upwelling into the tent that could be measured, but it would also be helpful to record nitrogen, argon, and helium flows. With a statistically useful set of tents erected in an area, it would be possible to establish the approximate volume of the outflow of gases from below.

The advantage of this type of data collection is that the equipment costs would be small. Other, more expensive techniques would give results of greater consequence, however. One such technique would entail injecting a burst of a tracer gas into the ground at some depth. (An appropriate tracer gas would be one that does not appear naturally in the ground and that is chemically inert relative to the material of the rocks.) After the injection, we would observe the time elapsed before this gas is detected at the surface surrounding the injection point. Because this tracer gas must be carried as a small admixture of the gases that are normally streaming through the ground, the proposed measurement (together with knowledge of the porosity in the area down to the depth of injection) would yield the flow speed and quantity of gases upwelling naturally.

To perform this tracer gas experiment, we might drill a small-diameter borehole to a depth of a kilometer or so and cement a casing into it just short of the bottom, sealing off the lowermost segment. A small pipe penetrating the closed segment of casing would be used to inject the tracer gas in individual bursts. The only escape route for the tracer gas would be out the bottom and into the natural porosity of the rock. Measuring instruments for the tracer gas would then be deployed on the surface surrounding the well, and the time elapsed between the injection and the first appearance of the tracer at the surface could thus be used to calculate the upward speed of flow of the natural streams.

This tracer gas procedure would provide good data for projecting flow rates over a relatively wide area if the ground were homogeneous. The use of several wells would allow us to detect any major inhomo-geneities. The procedure would also yield another result, possibly of great interest to chemical surface prospecting: It would show whether the gases take a vertical path through the rocks or lateral pressure dif-ferences cause them to take paths that are inclined at some angle to the vertical. We would simply note whether the bulk of the tracer gas emerges at a point displaced from the injection point, rather than encir-cling the well smoothly.

This tracer gas technique to measure the volume and rate of hydro-carbon flow from depth into the atmosphere, along with the cruder tent method, could be applied in any number of gas fields in which hydro-carbon reserves are well defined. I believe that the results would fur-ther support the deep-earth gas theory and cast further doubt on the biogenic theory.

Microbial Investigations

Once the primordial nature of hydrocarbons and their upwelling from great depth is deemed plausible, it is yet another step to take seriously the possible existence of a huge and independent realm of life beneath our own surface biosphere. But indigenous microbes have, in fact, been drawn up in hydrocarbon fluids encountered at great depth in oil wells, as described in Chapters 2 and 6. And all liq-uid hydrocarbons bear the signature of organic molecules of a kind and occurrence best explained by the posited existence of a planetwide horizon of subsurface microbes feeding on this rich source of chemical energy (Chapter 5). How do we begin to learn about the inhabitants of the deep hot biosphere?

It is difficult enough to study microbes that are comfortable with surface conditions and to ascertain food web relationships among the living and their metabolic products. It is very difficult to do these things for microbial ecosystems that humans will never be able to visit directly and whose inhabitants are accustomed to temperatures and pressures that differ greatly from the ambient conditions in our surface laboratories (and would be extremely expensive to duplicate). How, then, can we be sure which metabolic activity is primary—that is, which is the foundation of the food web? How can we determine what is the original source of energy that drives the whole system?

It is easy to determine the base of the food web and the original energy source for any ecosystem of the surface biosphere. This is because all such food webs ultimately depend on photosynthesis. Find the photosynthesizers (or the buried remains of the photosynthesizers), and you find the base of the food web. But finding the foundation of a chemically based ecosystem can be difficult in the extreme. How can one determine whether a particular chemical constituent is an original resource or a biological product?

From our understanding of life in the surface biosphere, we know that colonies of microorganisms will develop in any location where the energy, material, aqueous, thermal, and chemical requirements of life can be satisfied. Photosynthesizers at the base of the surface food web convert the energy of sunlight into chemical energy, from which they fuel their metabolisms and build their bodies. Proceeding up the scale of consumption, whatever energy is left over in the bodies or metabolic products of one group is tackled by the next group in the sequence.

At deep-ocean vents, too, we can expect to find a number of different chemical processes all mixed up in the same body of water. This is because everybody's waste is likely to be somebody else's food. The hard part is to decide which is the primary step that supplies the energy for all this biological activity and where the essential nutrient chemicals come from. Reduced molecules that can be oxidized will supply the energy. The same molecules, supplemented by other varieties, will supply carbon, hydrogen, water, oxygen, nitrogen, and (in smaller quantities) various other chemicals that are required to initiate complex biochemical reactions. To determine empirically which substance is the primary energy source and which is the oxidant for a chemically based system requires more information than can be obtained from studying the metabolism of any one type of inhabitant.

To study microbial life at depth in the earth's outer crust or at a borderland site, such as deep-ocean vents, it is customary to collect from the site a sample of biologically active fluid (inoculant) and then attempt to culture this sample in a vessel that has been sterilized and provided with the material conditions hypothesized to be necessary for growth of the indigenous microbes. For example, if hydrogen sulfide is provided in the cultured sample, and molecular oxygen is also introduced, then any growth that ensues can be assumed to have used hydrogen sulfide as the fuel source and molecular oxygen as the oxidant. If carbon dioxide is provided for the carbon source, then bacterial growth under those conditions indicates the presence of carbon-fixing microbes.

It is crucial to remember that in culturing microbes, if you do not provision a growth chamber with methane to serve as fuel, you will not discover the presence of methanotrophs. If you do not attempt to nourish a bit of inoculant with a carbon source such as methane, rather than carbon dioxide, then of course you will not discover an organism that utilizes methane and not carbon dioxide. Because metabolic studies of deep-ocean vent microbes have detected biological activity on a laboratory culture medium of hydrogen sulfide, oxygen, and carbon dioxide, it is generally assumed that the organisms that use these materials represent the base of the food web. But one of the key participants in this research, David Karl, urges his colleagues to "keep an open mind."[2] The mere fact that one living system has been verified does not mean that this is the primary one and that there are no others.

Accordingly, I hope that more attention will be given to the source and role of methane in this ecosystem. As noted in Chapter 2, microbes that derive their energy and their carbon from methane (methanotrophs) have been discovered in symbioses with mussels at the deep-ocean vents.[3] I anticipate that continued research will eventually reveal that methane is an important chemical foundation of the deep-ocean vent ecosystem. It is equally possible that hydrogen sulfide will prove to be a waste product of methanotrophs that obtain oxygen from sulfate, in which case methane will be awarded the foundational position.

Introducing the correct fuel source, oxidant, and nutrients into the laboratory culturing vessel is just the first step. It is equally important to provide the inoculant microbes a temperature and pressure bath to their liking. If a sample is drawn up from an oil well at 110°C and 600 atmospheres pressure, then those same conditions ought to be made available to the microbes in their new home. It is widely known in the microbial research community that hyperthermophiles cannot be expected to grow at temperatures that humans find comfortable, and high temperatures are not difficult to provide in the laboratory. But it is another matter entirely to offer microbes 600 atmospheres of pressure. Few, if any, microbiology labs are so equipped. Lack of sufficient pressurization in culturing experiments may be especially detrimental for inoculants that contain methanotrophs drawn up from great depths. Methanotrophs may be unable to access the diffuse vapors of methane at atmospheric pressures. For them, depth (or simulated depth) may be not just desirable but essential.

Until we furnish our laboratories with culturing vessels that mimic the conditions of intense heat and pressure characteristic of great depths, we should not interpret an absence of microbial activity in culturing experiments as evidence of the absence of life at such depths. Furthermore, if biological molecules are identified in a sample but cannot be coaxed into metabolic activity, we should not automatically discount those molecules as contamination introduced from the surface while the hole was being drilled. Instead we may have inadvertently killed the microbes in transit to the lab.

Even if ideal laboratory conditions are available, successful culturing of deep-earth microbes may be undercut by the methods used to transport biological samples to the lab. High temperatures and high pressures may not only be required for growth; they may be required for survival. Exposure to atmospheric pressure for even brief periods might prove lethal, just as some fishes reeled in from depth cannot be expected to live and return to their home when the hook is removed at shipside. John Postgate, in his book *The Outer Reaches of Life*, tells the story of researchers attempting to culture a sample drawn from a site in Antarctica.[4] The first culturing experiments yielded microbes that could function and reproduce at very cold temperatures. But these microbes were not absolutely dependent on cold temperatures; they could reproduce at warmer temperatures too. Not until samples were kept in frigid conditions throughout the entire journey from field to laboratory were new microbes discovered that could survive only in cold temperatures.

This story is a cautionary tale for researchers attempting to culture samples from deep wells. Unless samples from deep wells are provided with suitable conditions all the way up the tube and en route to the laboratory, we cannot expect the life forms to arrive in a state that will permit us to waken them from their repose. How difficult may it be to provide suitable conditions? If a sample were taken at a depth of 5 kilometers, the pressure there would be anywhere between 500 and 1500 atmospheres. It would take very strong vessels indeed, and exceptionally good seals, to yield samples of a biological integrity representative of such depths. In turn, any culturing experiments must be done at pressures and temperatures similar to those at the origin of the samples. If the pressure down there were anything approaching 1500 atmospheres and the temperature 120°C, the culturing apparatus would be costly to build and operate. No reports of such devices have come to my attention.

There is, however, another and perhaps much simpler way to proceed. We might send down the borehole a container of mechanical strength sufficient to resist being crushed by the fluid pressure. The container should have two chambers that can be opened and closed by command from the surface. One chamber could be furnished with a common bacterial culturing medium, such as acetate or sugar. The other would be provisioned with the material thought to be the oxygen donor for microbial life down there. At the chosen depth, we would open both chambers and let them fill with the drilling water and then close them again. The interiors would thus be maintained at the pressure and temperature at which the sample was taken. The container might be left at that depth for the time it would take for cultures to develop. When the experimental chambers were hauled back up into the low-pressure and low-temperature regime of the surface biosphere, much of the microbial material might be killed. But the end products of their metabolism could still be discovered. Thus if one chamber had been provisioned with the oxidant ferric iron in the form of small particles, but on return to the surface it was found to contain the iron in a less oxidized state, such as magnetite, we would conclude that the inoculant introduced at depth had carried an infectant whose metabolism required the reduction of ferric iron, possibly using the reduced gases or liquids that had accumulated in the wellbore.

By these and other innovative culturing techniques, the entire ecology of deep life may one day be revealed. Step by step, the various microbes of the deep will be sampled, cultured, and understood. Some interpreters of the results will undoubtedly still conclude that our own surface biosphere has mastered strategies for surviving in an extreme environment, feasting on the geologically processed remains of surface life long dead and long buried in the sediments. Others may conclude, with me, that we have come upon an entirely new world.

Prospects for Extraterrestrial Surface Life

Life as we know it depends utterly on the presence of liquid water, as was noted in Chapter 1. Ice or steam will not do. Water in its solid form loses any ability to serve as a matrix for the mixing and matching of molecules. If nothing can move, no recombinations of atoms can take place. If the temperature is more than a few degrees above the local boiling point of water, water escapes as steam from any biological material. Water vapor mixed in an atmosphere of other gases can, however, supply water to biological systems and can even be turned into liquid water by chemical action. But this can happen only in the temperature range between freezing and boiling.

If water—and water in liquid form—is indeed a requirement for life on the surface of a planet, then a combination of several conditions necessary for water's presence must exist, notably the nature of the central star, the distance of the planet's orbit from that star, the size and mass of the planet, and the nature of the planet's atmosphere. If photosynthesis is required for the main energy supply, then a substantial temperature difference between the surface of the star and that of the planet is also required. Light of suitable wavelength and the presence of liquid water together thus dictate what turns out to be a very narrow window for surface life.

Water is a common molecule in the universe, but water in its liquid phase is rare on planetary and satellite surfaces of this solar system. There is only one body so blessed: our own blue pearl of a planet. We need look no farther than our two nearest planetary neighbors, Venus and Mars, to realize just how lucky it was for surface life that the earth turned up in the right place, was of the right size and the right composition, and had the right kind of star to illuminate it.

Let us begin with Venus. Venus receives almost twice as much solar radiation as does the earth. Climate model calculations[5] indicate that if the earth had to bear that level of solar influx, a "runaway greenhouse" would ensue, destroying all prospects for surface life. Indeed, because the sun is slowly heating up, a runaway greenhouse is in the earth's future, too—but not for billions of years.[6] A runaway greenhouse works in this way: First, higher radiation intensity means a hotter climate to begin with, which vaporizes more liquid water on a planet's surface. The hotter climate also permits more of that water vapor to linger in the atmosphere before falling out as rain. Water vapor is a greenhouse gas—in fact, it is the most effective greenhouse gas in the earth's atmosphere. It is transparent to most of the incoming solar energy, but when some visible light is transformed into the infrared at the earth's surface, radiation of some of this heat back out to space is blocked. Water molecules present in the atmosphere as vapor reflect these longer wavelengths back to the earth's surface, further heating the planet and enabling the atmosphere to hold even more water in a vapor state—which feeds back on the amount of heat retained by the earth, and so on. Eventually, all surface water would vaporize.

Worse yet, an intense hothouse climate would push a lot of that water vapor into the upper reaches of the atmosphere, where its molecular bonds would be vulnerable to the most energetic wavelengths of solar radiation. When molecules of water are broken up into hydrogen and oxygen, the light-weight hydrogen component (H_2) escapes from the upper atmosphere directly into space. The remaining oxygen atoms may then either join the store of atmospheric oxygen or increase the degree of oxidation of the planet's surface materials. Any carbon is likely to be oxidized to carbon dioxide, which is fated to remain in the atmosphere of a waterless planet. CO_2 is a potent greenhouse gas, and it will be stably confined by a gravitational field as strong as that of the earth or of Venus. So long as the planet continues to drive out carbon-bearing gases from its interior, CO_2 will accumulate in the atmosphere.

The runaway greenhouse on Venus has advanced to a waterless end. Although Venus was formed of the same solar system debris as the earth, and probably also started with some water content, no water vapor can be detected in its atmosphere today. The Venusian atmosphere is dominated by CO_2.

On our own water-rich world, the carbon atoms have been removed from the atmosphere at about the same rate at which they were supplied by outgassing from the depths of the earth. Methane and carbon dioxide are the principal sources of carbon entering today's atmosphere, as noted in earlier chapters, and methane also turns quickly into carbon dioxide and water in our oxidizing atmosphere. The statistics of boreholes and the investigation of meteoritic materials (the leftovers of the construction of the planets) both strongly suggest that methane is the major gas involved. Carbon dioxide is withdrawn again from the atmosphere, permanently or on a long-term basis, mainly by being sequestered in carbonate rocks deposited on the ocean bottom. Interestingly enough, had this water-mediated route for carbon dioxide removal not been functioning on the earth, our own atmosphere would now hold just about as much carbon dioxide as does that of our sister planet Venus, where the surface atmospheric pressure is about eighty times ours. There would be no oceans, no rain, and no surface biosphere.

Turning to Mars, we encounter the opposite problem for water-dependent surface life. Whereas the intensity of solar radiation on Venus is twice that on the earth, the intensity on Mars is less than one-half of ours. Not surprisingly, Mars has a surface that is everywhere below the freezing temperature of water, and in some areas and seasons, it is very far below that. It seems that whereas Venus is too close to the sun, Mars is too distant. But Mars has an even bigger disadvantage than its remoteness. Mars is too small.

Surface temperature depends both on distance from the sun and on the green-house capacities and density of an atmosphere. The smaller the mass of the planet, the less massive will be the outgassing process that creates an atmosphere, and the weaker the gravitational force that acts to compress it. Such an atmosphere will also be particularly vulnerable to solar outbursts of the intensity seen a few times in every eleven-year solar cycle. These outbursts can sweep away a bit of atmosphere from a low-gravity planet or moon, especially if the body does not have a magnetic field to shield it from high-energy ionized particles emitted by the sun. Because Mars has only a tenth of the mass of the earth (a third of our gravity) and no general magnetic field, it would easily suffer such losses.

Today's Martian atmosphere exerts only seven-thousandths the atmospheric pressure of our own. It is almost entirely composed of the greenhouse gas carbon dioxide, but there is far too little CO_2 to raise surface temperatures enough for ice to melt and rain to fall. Indeed, the poles are so cold that the ice caps are made partly of carbon dioxide ice.

The small mass of the atmosphere of Mars narrows the possibilities for liquid water in a more important way. Anyone who has ever tried to cook at high elevations will be familiar with this problem. At high altitudes, there is less atmosphere pressing down on the surface of a pot of water. Liquid water therefore vaporizes more readily, and the boiling point declines. This means that the atmospheric pressure on a planet defines the temperature range in which liquid water can exist on the surface.

At sea level on the earth, the atmosphere pushes down with a weight of about 1 kilogram per square centimeter of surface area. That pressure provides a 100°C window for liquid water, between boiling and freezing, which means a 100°C window for any forms of surface life to engage in metabolic activities. (This range can be increased a few degrees by chemicals dissolved in the water). A range of 100°C for liquid water is not much more than is needed for life to cope with the range of solar heat delivered to different parts of our own planet. On Mars this window for liquid water is much narrower, for although the boiling point of water is critically dependent on atmospheric pressure, that pressure hardly affects its freezing point. The atmospheric pressure that the present small mass of the Martian atmosphere exerts would not allow any liquid water to exist there; instead it would freeze and sublimate (pass directly from solid to vapor without first passing through the liquid state). To have as great an atmospheric pressure as we have here on the earth, and therefore to have a similar temperature range for liquid water, would require, in the weak gravitational field of Mars, three times as much atmospheric mass above each square centimeter of surface. There would have to be more than 140 times as much mass as there is in the present Martian atmosphere.

But can one assume that such an atmosphere ever existed on Mars? Where would it all have come from and how would it have disappeared? CO_2 is a heavy molecule (44 atomic mass units) and would be held firmly to Mars, whereas the light water vapor molecule (only 18 mass units) could escape or suffer dissociation by high-energy components of sunlight, allowing the hydrogen then to escape into space and the oxygen to remain on the planet, either in the atmosphere or fixed in surface materials. The same process happens on the earth; but here, with greater gravitational pressure and with the shield of a magnetic field, little water vapor escapes directly into space. Some dissociated hydrogen escapes though, as noted earlier, and the dissociated oxygen joins the large pool of atmospheric oxygen.

Was there ever liquid water on Mars? Could the Martian surface once have supported life? It has been widely assumed that rivers did indeed flow on Mars in the distant past. The images beamed back by the Viking spacecraft in 1976 seemed to offer evidence of a formerly well-watered world. To this way of thinking, dry channels of once-enormous rivers are prominent features of the Martian landscape. Some channels are so large that they are thought to have been gouged by sudden ruptures of natural dams, releasing torrents of water far greater than have ever rushed across the earth.

I (and others before me[7]) believe that the surface flow patterns on Mars can be explained much more satisfactorily by glaciation. To my eye, the very broad and smoothly curving valleys dubbed "superhigh-ways" have shapes quite characteristic of glacial flow and not at all like the flow of a liquid. For example, where a major obstruction lies in a channel, the flow pattern splits around it and recombines behind, cre-ating a teardrop-shaped island. All directional changes of these chan-nels are gentle; the curves do not encompass more than 30 degrees of directional change in 100 kilometers. Channel edges are smooth with no sharp indentations, and the banks maintain the same slope for hun-dreds of kilometers. Such features cannot be explained by water, how-ever fast it might be asked to flow. Fast-flowing water is turbulent; the direction of flow is maintained only for short distances before turbu-lence, and even slight irregularities of terrain, substantially redirect the river. In contrast, ice is a solid, or, to be more precise, a liquid of enor-mously high viscosity. It is this viscosity that makes the flow smooth, slow, and free of turbulence. The stiffness of the material opposes any flow patterns that would force abrupt changes in direction.

Although the "superhighway" channels are best explained as recording movements of earlier vast glaciation on the surface of Mars, there are other channels that show all the curves and unevenness of ter-restrial rivers and that are marked by abrupt changes in direction. Some even show the kinds of loops that we see on the earth's great river deltas. Such features certainly suggest that liquid water once flowed there, but they are now dry rills in the Martian surface. Would this imply that a huge atmosphere once protected liquid water on the surface and that the atmospheric temperature was above the freezing point?

I do not think so. Again, the existence of thick glaciation in earlier times would account for these features. Flowing water is, after all, common under terrestrial glaciers, and this water digs out the flow patterns of rivers in the ground below. On the earth we see fluvial features in many cases where a glacier has retreated. The same process can be expected to have happened on Mars. If the ice sheet had been 2 kilometers thick, it would have exerted a pressure in the gravity field of Mars equivalent to 67 earth atmospheres. A thick ice sheet would also have provided heat insulation so that water warmed from deeper levels (by the same processes of gravitational sorting, compression, and radioactive decay that heat the earth's interior) could penetrate and flow beneath the ice. Such subglacial rivers would have carved channels and transported material just as they have done on our own planet.

In the course of time, the Martian glaciers would have sublimated into vapor. Because of the weak atmospheric pressure, glacial ice would not have needed to go through a liquid stage. Sublimation, not melting, would thus have put an end to the Ice Age on Mars. Channels carved by glacial flow and deposits made by subglacial rivers of liquid water would have been exposed relatively intact. A multitude of rocks of various sizes and types that were scattered by large impacts onto the different levels of the glaciers during their slow evaporation would eventually all settle gently onto dry surface, explaining the origin of the numerous fields densely populated by a diversity of rocks.

One of the very first Viking photos of 1976 shows, resting on the surface, a rock that is much larger than all others in the dense field. Assembled around it are much smaller rocks that clearly define good approximations to circular patterns. Just this type of feature has often been observed (by me and by many others) on terrestrial glaciers. What accounts for this pattern? The large rock will heat up during the day, retaining a lot of heat that will eventually be radiated out. The surrounding ice, which captured less heat in the first place, will be warmed a little by that rock and therefore will evaporate a little faster than the other ice. A funnel in the ice will develop around the rock. Whatever pebbles are locked in the ice will tumble to the bottom of the funnel as it forms. If the temperature and evaporation rates vary in time (as has certainly been the case on the earth), rings of debris will come to surround the large rock.

Figure 10.1 A small glacial feature on Mars. Rings of rocks
encircling the large rock at the center give strong evidence of glaciation and subsequent evaporation of ice. Many similar features have
been seen on terrestrial glaciers. Courtesy of NASA.

Photographs of Mars obtained by NASA's Pathfinder mission in 1997,
and by the remotely controlled vehicle roving the planetary surface, confirmed the findings of the Viking missions two decades earlier. Densely
scattered rocks of quite different compositions are a common feature, and
a powdery rock material occurs in the spaces between rocks. In my view,
the density of rocks in these fields is such that it would be difficult to
account for them as objects thrown out at high velocity from impact
craters. Very little destruction is seen in these rocks. In contrast, a strewn
field from impacts would generate breakage, because many of the trajec-
tories would be at a small angle to the surface and would thus smash
many other rocks before coming to rest. A glacial explanation better
explains the micro as well as the macro features detected on Mars.

One more large-scale feature on Mars that glaciation would explain
is the prominent escarpment that surrounds the base of the largest
mountain, Olympus Mons, which is possibly the largest volcano[8] in

the solar system. This very steep escarpment is between 1 and 2 kilometers high, all around the mountain. It is very difficult to explain how lava stopped flowing down the mountain exactly to this edge or how such a large quantity of frozen lava could have been eroded away subsequently. But if the volcano had been surrounded by glaciers at the time of the last large eruptions, and if the ice had been 2 kilometers thick, then whatever liquids came down the mountain would have been arrested at or near the glacier level. Subsequent sublimation would have left the steep escarpment. Similar features have been reported on terrestrial volcanoes that border on glaciers.

These and many other features of Mars point to heavy glaciation, rather than to vast, surface-flowing rivers. Interpreting such features as glacial also solves a major problem that planetary scientists encounter when they try to construct models of the climate history of Mars. It is very difficult to propose plausible scenarios by which Mars could at one time have retained an atmosphere massive enough to hold down liquid water that then would have to disappear, leaving less than a hundredth of its original mass behind—all while carving a "fluvial" landscape by means of floods far greater than those that have coursed over the surface of the earth.

If Venus is too hot to support water-based life on its surface, Mars, then, is too cold. The rocky moons of the large outer planets are even colder than Mars, but unlike Mars, these moons are so cold that frozen waters on their surfaces survive for long geological periods. Surface life on other planetary bodies in our solar system thus seems to be excluded. *Subsurface* life, however, is another matter. Mars, the satellites of the major planets, many asteroids, and even our own moon should be regarded as real prospects for harboring extraterrestrial life of this kind.

Deepening the Search
for Extraterrestrial Life

A most exciting field of study in the future, and one in which we will learn much about the origin and evolution of life, will be the investigation of subsurface life inhabiting other planetary bodies in our solar system. It would be a great help, in these necessarily expensive and prolonged exploratory missions, to have gained already a clear understanding of the first such example: the earth's deep hot biosphere. Close cooperation between the space program and deep geochemical investigations of our own planet could be mutually beneficial, allowing discoveries to be made that are important to both efforts.

Although the surface conditions on the other solid planetary bodies are all quite different from those on the earth, the subsurface conditions within many of the larger planetary bodies will be similar to ours. The relationship of pressure and temperature with depth will, of course, be different, but the chances of life having developed at some depth may be not too different from those here. Hydrocarbons (methane and others) have been spectroscopically detected on the surfaces and in the atmospheres of many of these bodies, and subsurface liquid water can be expected within most of them (water appears to have been plentiful in the circumsolar cloud that formed the planets, and ice has been identified on several other planetary bodies and comets that are colder than the earth). The rocks, like those of the earth, should contain some oxidized components that will serve as oxygen donors. The scene would thus be set for the existence of microbiology there.

Mars would be the least expensive planetary body to investigate for evidence of subsurface extraterrestrial life, because we might not need to launch any spacecraft to begin such an effort. Meteorites that occasionally fall to earth bear the chemical signature of Mars. Several meteorites collected from the ice fields of Antarctica appear to have come from Mars. Trace element ratios such as the sequence of noble gases from neon to xenon, as well as the rather unusual nitrogen isotope ratio of the Martian atmosphere, were measured by the earlier Viking landing craft, and very similar values for these quantities show up in these meteorites. It seems very unlikely that debris from any other body would match these quantities so closely. Millions of years after an impact on Mars caused the ejection of Martian material, the orbits of some of these ejecta led to collisions with the earth.

In 1996, one such meteorite (denoted ALH84001) yielded strong evidence that the rock had been altered by microbial life while it was still part of Mars.[9] Detailed examination made it seem very improbable that this evidence was due to contamination from life on the earth. Rather, the biological imprint had been present in the interior of the stone before it was ejected from Mars. Like many meteorites, ALH84001 was found on the surface expanses of glacial ice in Antarctica free from any terrestrial debris.

Meteorite ALH84001 gives emphasis to our search for extraterrestrial life because the rock almost certainly came not from the Martian surface but from some depth. For an object to be shot from Mars into an orbit that could eventually end on the earth, a very large impact on Mars would have to have been responsible. There are many large impact craters on Mars, so this does not seem improbable. In a large impact, most of the material excavated, and possibly propelled to a high velocity, will have come from a considerable depth. If that is the case, evidence in the meteorite of biological processes is evidence of life at depth—not of surface life that was later buried in new surface Martian sediments.

In a small meteorite on a long flight through space, the liquid or gaseous products of the deep microbiology should generally escape in a short time. It is the solids created by life that give the clue. What solids should we look for? If carbon-based life within Mars feeds (or once fed) on primordial hydrocarbons upwelling from a greater depth, and if Martian life required oxygen in order to access the energy offered by the chemical reaction of oxygen with the hydrocarbon molecules, then one piece of evidence would be the altered state of the oxygen donor. If iron oxides served as the oxygen donor, then the end product would be iron in a less oxidized state, which happens to be magnetic. Magnetite is the most common form, and as noted in Chapter 6, the abundance of exceedingly small particles of magnetite discovered at depth in our borehole in Sweden was important evidence of life at work. Alternatively, if sulfur oxides were the oxygen donors, we would expect to see metal sulfides as the permanent remains. The product of the oxidation of the hydrocarbons would be carbon dioxide and water, and in many rocks these would react with oxides of calcium or magnesium to make solid carbonates. In our own deep biosphere, the carbonate by-products of life take the form of cements that fill up small pore spaces.

We know that highly oxidized iron is abundant on Mars, and very small-grained magnetite has in fact been detected in the Martian meteorite that has been studied for signs of biological alteration. That meteorite also contains iron sulfide and carbonate cements. Moreover, it contains polycyclic aromatic hydrocarbons, which could be the large molecules that remain of a primordial crude oil once contained in the rock, whose lighter molecules vanished after many millennia of exposure to the vacuum of space. In addition, the meteorite contains small objects, detectable under scanning electron microscopy, that may well be fossils of microbes. Although the purported fossils by themselves would not be conclusive evidence, the association of magnetite, iron sulfide, carbonates, and heavy hydrocarbons, to my mind, makes a strong case for the microbial explanation. It is true that each item can be produced without biological intervention, but the odds of finding them together accidentally in a small volume would be very low. Many terrestrial oil and gas wells show just such an association (but an association with helium also, which the meteorite could not have transported through space).

The case for subsurface life dwelling within Mars is now so tantalizing that a visit to Mars may be scientifically very productive on this question. For future interplanetary missions that could return a sample to the earth, it would be best to go to locations where material is now exposed that must once have been at some depth.[10] The floor of the deep "Valles Marineris" is such a place. There, massive landslides have exposed material that must once have been at a depth well into the liquid water domain. Large impacts will also have dug deep into the water domain, so samples that have been tossed up in such events may also show water-related effects.

In my 1992 paper presenting the deep hot biosphere idea, I suggested that perhaps 10 planetary bodies in our solar system would provide suitable subsurface homes for fundamentally the same kind of life as we have within the earth.[11] I made that prediction by using a rather simple and generic formula: Any rocky planetary body at least as big as the earth's moon might be expected to offer the requisite subsurface conditions of heat and upwelling hydrocarbons.

Two considerations led me to choose our moon as the lower size limit for a subsurface biosphere. First, bodies smaller than our moon will probably fail to support liquid water even at depth. Second, empirical evidence indicated the presence of lunar hydrocarbons. The only conceivable cause of the deep, internal quakes that have been observed on the tectonically frozen moon is the opening and closing of pores as fluids upwell. There, as with the earth, those fluids should contain primordial hydrocarbons. An instrument placed on the moon by an Apollo mission detected gas particles of mass 16 (in atomic mass units) at the times of such quakes. I do not know of any atom or molecule of that mass that would be stable and unreactive enough to have made its way through the lunar rocks, other than methane.

In 1992 I thus concluded that subsurface life was indeed possible within a number of planetary bodies. But I could not then judge whether extraterrestrial life was not just possible but probable—perhaps even common. Since then I have become convinced that life is highly probable within those bodies. I have no more substantial empirical evidence now than I had then. The argument, rather, is based on logic. To begin with, I assume not only that the deep biosphere within the earth is independent of surface life but also that it was the progenitor of surface life. If I am correct in concluding that 10 planetary bodies offer subsurface conditions suitable for life, and that our own planet alone is suitable for surface life, then it would be highly improbable that a deep biosphere happened to develop only on the one planet that could also support surface life.

Independent Beginnings or Panspermia?

If subsurface life does exist elsewhere in this solar system, would it have originated independently on each planetary body? Or might there be natural mechanisms by which life is transported from one planet or moon to another? This latter mode of interplanetary biological infection has come to be called *panspermia*.

To answer these questions, we would need to consider whether microbial life might evolve spontaneously in all locations that are favorable. What is the likelihood that life of independent origin would have adopted a similar basic chemistry? And if interplanetary infection were a mechanism, how would it work?

If on another planetary body we were to find a type of biology that used quite different basic steps of chemistry, outside the range of the variants we have observed for earth life, we would conclude (though not with complete certainty) that it represented an independent origin. That discovery would suggest, in turn, that some variants of life should arise with high probability in many other favorable locations. But if we saw life forms with a similar basic chemistry, could we then make a distinction between panspermia and a very closely parallel evolution? Perhaps our own biochemistry is the only one that could make functional organisms, in which case no other would be found. Or perhaps ours is one of a small number of possible biochemistries and, for this reason, would probably be discovered elsewhere.

What if the biochemistry of an extraterrestrial ecology proved to be nearly the same as that of the earth—but of the opposite chirality? As in Chapter 9, a molecule is chiral if there can be another molecule that is identical to the mirror image of the former but differs in no other way. DNA, for example, is a helical molecule that spirals to the right. Its chiral opposite would be the same molecule spiraling to the left.

If we found, in biological molecules of another planetary body, the same basic chemistry as the one we have here, but the chirality was opposite of ours, we would have substantial evidence to conclude that life, using the same basic chemistry, had a good probability of arising *independently* on other bodies that had similar subsurface conditions. If, however, we found the same chirality there, all we could say would be that they might derive from an evolution in common with ours (panspermia), or that an independent origin favoring the same basic chemistry might have hit (with a 50–50 chance) on the same chirality as ours.

If we repeated such observations for yet another planetary body and thereby obtained a second extraterrestrial example with the same, earth-like chirality, we would conclude that the evidence was beginning to point toward a common origin, because an independent origin would have had a probability of only 1 in 4 of providing the same chirality in three independent cases. The investigations of yet more planetary bodies would then become essential for resolving the issue.

By what mechanism might life that originated within one planetary body spread to another? The meteorite from Mars that was found in Antarctica provides the answer: a massive impact that splits off pieces of a planet or moon and ejects them into space. In addition to the trauma of exit from the home planet and entry into a new one, the life-bearing rock would be subjected to cosmic rays while in transit. If the journey were long, the vicissitudes of extended dormancy would be another challenge for successful infection of a new planetary body. Damaging radiations and the passage of time would become especially severe constraints on the natural transport of life between star systems—galactic panspermia. The odds would improve if there were bodies of planetary size that could support subsurface biospheres in the space between stars. The largest sophisticated telescopes would have great difficulties detecting such bodies.

Molecular clouds might well be forming such dark objects constantly, and only a fraction would come to be associated with a star. They could contain, and maintain for billions of years, an active internal microbial life, just as seems to be the case for the earth. Accordingly, they could carry active life at depth through space over interstellar distances. When such an object came, perchance, into the vicinity of a planetary system, collisions with planets would then allow active, living material to be distributed. Impacts with smaller objects that spall off pieces like the Martian meteorite could achieve the same result. In this case there would be no dependence on dormant life for long periods, nor on any long-term resistance to damage from cosmic rays, two problems that have made other galactic-scale panspermia proposals seem improbable. Even though the odds against the galactic panspermia described here are very high, it is not impossible. Panspermia between planets of one planetary system would clearly be a possibility, as the Martian meteorite has shown.

Is all this just idle speculation? Should we take no notice of the views presented in these chapters? Should I and others discontinue investigating the deep hot biosphere and probing for microbial life beneath the earth's surface? Should the scientific community also discourage systematic researches into the origin of petroleum, the reason for the strong association of helium with petroleum, the cause of earthquakes, and why different metal minerals occur together so frequently in the same location? These are all features that have yet to be explained by the reigning theories, so particular attention is due them (rather than the neglect that inexplicable phenomena tend to receive in modern scientific literature). The history of science offers example after example of apparently inexplicable features for which perfectly rational explanations were finally found. In nearly all such cases, assumptions that were universally believed obscured the truth so effectively that no progress toward a solution seemed possible. Yet it is to just such apparently inexplicable features that we must hope to find clues that will show us how to unburden ourselves of false beliefs.

Speculation is a vital step in this process. It was once speculated that the earth revolved around the sun. Without this speculation, I do not think that any systematic avenue of research would have produced the evidence that clinched this theory. At a time when proposed solutions are still speculative, they are the driving force for the researches that will prove them right or wrong and will thereby put our thinking on a new and better track.

Afterword to the Paperback Edition

S ince the first publication of this book, in December 1998, I have only become more convinced that petroleum and black coal are not fossil residues that have worked their way down from the surface of Earth into their subterranean resting places. This widely accepted view of their *biogenic* origin is, in my view, mistaken, and this book proposes an alternative—namely, that Earth's massive reserves of hydrocarbons are *abiogenic,* that they were part of the primordial "soup" from which our planet was created, and that to this day they exist in abundance deep within our planet and continue to upwell toward the surface.

My reasons for holding this admittedly controversial view are numerous. First, it has become quite evident to me that the *quantity* of black coal and petroleum (and especially its natural gas component, methane) are far greater than could be explained by any theory that depends on buried biological debris. Second, petroleum and methane have been found and continue to be found in *locations* on Earth to which surface biological remains have never had access; the presence of oil and gas at these sites simply cannot be explained by the biogenic theory. Third, one finds at these sites none of the other residues one would expect to find in the presence of biogenic hydrocarbons. And fourth and perhaps most tellingly, it is now generally agreed that there is a profuse supply of hydrocarbons on many other bodies in our Solar System, where no origin from surface biology can be suggested. Yet the use of the name "fossil fuels" for Earth's supply of hydrocabons is widespread, giving the impression that their origin in surface biology has been established beyond doubt. Any discovery that conflicts with this old interpretation is still often described as "most surprising" or "inexplicable," even though the evidence, in and of itself, is not questioned. This lack of connection between undisputed evidence and generally accepted theory is especially strange when one considers the widely held belief that there is a fundamental shortage of these so-called "fossil fuels." For many decades, as we have found more and more reserves of petroleum and other hydrocarbons, we have constantly had to revise our estimates upward. To me the alternate conclusion is inescapable: We are just not running out of natural gas, oil, and coal, and after intense usage over a century, we now know of more reserves than had ever been predicted in the past.

Any terrestrial hydrocarbon contains molecules whose biological origin is beyond doubt, but this does not prove that the hydrocarbon's *origin* is biological. The alternative solution, set forth in this book, is that all petroleum we obtain from the ground has suffered a large amount of biological contamination at levels deeper than our drills can reach. Living material, deep beneath the surface of our planet, has left its mark on the oil, gas, and black coal that eventually comes to the surface.

"But the amounts of biological contaminants are very large," might go a reasonably skeptical counter-argument: "Where would there be space for all the living material required in your interpretation? And what would be its source of nourishment, its source of carbon and of the many other elements necessary to support life? And is it conceivable that we would have failed to become aware of the existence of such a large underground domain of life?"

If the evidence for the deep origin of petroleum is strong, as it appears to be, then a massive underground biology *has* to exist. This possibility had not been considered in the past, even though chemists who had studied the composition of petroleum had strongly hinted at it: "A primordial substance to which bio-products have been added," was the description of petroleum by Sir Robert Robinson, the 1947 Nobel

Laureate in Chemistry. Today, however, there is clear evidence for the existence of a massive deep biosphere. It represents not just a small perturbation in the scheme of geochemistry, if it is to account for all the biological substances that hydrocarbons are bringing up and have brought up over long periods of geologic time. It is instead vast. Space for all this living material is not a problem, provided it is microbial life, thriving in the pores of rocks. In the crust of Earth, where porosity of a few percent is common, and extending over a vertical dimension of perhaps 10 kilometers, this massive separate domain of life could amount to far more than the volume occupied by all surface life.

What about nourishment? Hydrocarbons seeping upward will supply the carbon, and can supply the energy for life if oxygen is available for their combustion. The great supply of oxygen in the surface atmosphere does not, of course, reach deep levels. And the oxygen found in subterranean rocks is too tightly bound to be useful for supporting life, as it would take more energy to liberate it than can be obtained from its use for the combustion of hydrocarbons. There are, however, a few common subterranean substances that can deliver oxygen sufficiently cheaply. The principal ones are oxidized sulfur in the form of sulfates, and highly oxidized iron, the residues of which are, respectively, sulfides and low-oxidation states of iron such as magnetite. And indeed, massive amounts of sulfides and magnetite are found just in petroleum-rich areas. Similarly, there are large number of cracks on the ocean floor that vent hydrocarbon gases, and these gases provide nourishment to profuse microbial life surrounding the vents. These sites are also surrounded by large deposits of metal sulfides, because in this case the oxygen donor appears to be the sulfate ion, a prominent component of seawater.

Other evidence for a constant supply of deep-source oil and gas comes from observations in the petroleum industry. We have seen oil and gas fields refilling themselves, sometimes as fast as they were being drained, and many fields have already produced several times as much as earlier estimates predicted. Petroleum scientists have also found that oil often contains trace elements quite different from what could be expected from the rock underlying a particular field. Nickel and vanadium have long been recognized in this category, but we must add to the list the inert gas helium, whose concentration has no other explanation than a deep origin, as well as a number of metals that are considered to come from much deeper levels—including iridium, gold, silver, and platinum.

In looking at the petroleum industry, one also has to consider the widespread and unexpected locales in which drilling is now taking place. In Russia, a major project has been under way to establish the extent to which bedrock, rather than sediments, contain hydrocarbons. More than 300 deep holes have been drilled in Tatarstan (in Central Russia), all into broken-up basement rock, to depths of 5 kilometers or more. The majority showed the presence of high levels of hydrocarbons. About a hundred wells in other parts of the world exist that were also drilled in basement rock, and many are producing petroleum. Among those, the recent discovery of a major off-shore site in Vietnam, the White Tiger field, is of particular interest, as very good oil production, largely from basement rock, is in progress there.

Then there is the evidence from the sea. Methane hydrate, an ice made up of a water–methane combination, covers very large areas of ocean floor, and the total quantities of the element carbon contained in this substance is estimated to be greater than the carbon contained in all coal and oil that has been identified the world over. This concentration of carbon could not have arisen from surface plant material sinking down, as such a supply is far too small and would have brought much other material down with it, which is not there. But beyond that, how could layers many meters thick composed of this solid ice have formed from a supply from above, when neither plant material nor bare methane gas could penetrate downward through the ice? Yet large bubbles of gas are detected, held down between the rock of the ocean floor and the ice layer above. This gas cannot have reached such locations except from cracks in the rocks below. Widespread methane outgassing is also indicated by land-based permafrost having similarly sealed-up methane hydrates as well as pockets of methane gas. This would imply that methane outgassing of the earth is a widespread or general process, quite in accord with the information we have from the ocean vents and their rapidly growing colonies of life.

Added to this are sudden outbursts of gas from the ocean floor. In some cases these violent phenomena have been directly observed, as in the eruption that caused a devastating tsunami on the coast of Papua, New Guinea, in July 1998. It has now been widely recognized that certain features of the ocean floor can be interpreted only as the result of the explosive eruptions of gas. Some have created circular depressions of as much as 200 meters in diameter, and have been found in many areas, including the East Coast of the United States, but also specifically in areas known to produce commercial quantities of natural gas, such as the North Sea and the East Coast of New Zealand. Measurements of gradual seepage of gas at quiet times have shown methane to be the principal component of these outbursts.

In this picture we can then understand that petroleum comes from deep levels in the Earth, from vastly larger reservoirs far below our reach, just as it must have done on the numerous other Solar System bodies that show large amounts of petroleum, but possess no surface life. On the way up, at levels that an oil prospector would call deep, but still much shallower than the origin of the petroleum, the temperature falls to levels at which some microbial life is possible, and where this life finds rich food supplies in these hydrocarbons and becomes prolific. This food supply is from chemical sources that Earth itself provides; it is not related to the photosynthesis of the surface. We are looking at an independent domain of life, not at an extension of the surface life we know. The quantities of microbial material in this biosphere can be estimated in two ways: one is from the microbial debris that remains in the hydrocarbon deposits, the other is from the solid mineral residue the microbial actions have left behind. Once these estimates are made, we arrive at quantities that are so large that it is now questionable whether by mass our surface biosphere is the principal domain of life on earth: it may be that this distinction belongs instead to the microbial mass in the pores of Earth's rocks.

The one connection between the two biospheres we see is the genetic one. The deep biosphere uses the same genetic processes and molecules as the surface biosphere, so presumably one derived from the other. But which came first?

Photosynthesis, which is the primary energy source for surface life, is a very complex—one might even say "fragile"—process. It cannot have stood at the beginning of life. Elaborate chemical processing must first have been invented, and the earlier life that gave rise to this processing must have had simpler chemical energy sources available to it. That alone would favor the deep biosphere for the beginning of life, but there are several other considerations that point the same way.

The most primitive forms of microbes, and therefore judged the earliest, belong to a distinctive branch. The physicist and microbiologist Carl Woese has shown that this branch is sufficiently different from bacteria that it should be given a its own classification, and has proposed the name "archea" for it. It has now been shown that the majority of thermophiles (heat-loving microbes) belong to this class, prompting the suggestion that life's origin was in warm or hot surroundings. This again favors the deep biosphere, which provides long periods of near-constant environmental conditions. In such an environment, an emerging life would not be interrupted by rapid changes, for example, in temperature, humidity, radiation, or winds, as would be the case on the surface.

Microbial life, or even smaller forms, would be the most likely candidate to take the first steps toward self-replicating complex life, if only on the grounds of probability. Microbial life displays by far the fastest adaptations, the fastest development of new features by Darwinian selection, because of its short cycles of reproduction and the large number of individual organisms in any one generation. For the same reason, we would also look for the beginnings of advanced life at the largest domain of microbial life we now know—and again, the deep biosphere fills the bill.

Much discussion of the origin of life has centered on the picture of a warm pool containing an assortment of elements or chemical compounds favorable for the creation of the chemistry of life. But there would be a problem. An essential aspect of life is reproduction, the re-creation of an existing form. Such a process must necessarily lead to an exponential increase of the numbers with time. If the first organism reproduces one like itself in a time T, then after a time $2T$ there will be 4, after $10T$ there will be 1024, and after $100T$ the result will be 10^{30}:

$$1,000,000,000,000,000,000,000,000,000,000$$

However large the pond may have been and however much "food" it may have contained, it would soon have been exhausted. The inevitable consequence is a brief feast followed by final famine. How does existing life avoid this, and go on for a long time? It does so only if there is the stabilizing effect of a food supply that is limited at any time, but constantly renewed. In our surface life we are so familiar with this, that we forget how essential it is. Here, on the surface, the sunlight is the limited and renewed energy supply; its energy is doled out each day, and it rises each morning (not forever, but for a very long time). The deep hot biosphere escapes the feast-and-famine situation by supporting life with the continuous, long-lived seepage of chemicals from sources that are too deep and hence too hot to be available to the life forms we have. So food supply at a "metered-out" rate exists there, just as it does on the surface.

All life is essentially an extension of the process of autocatalysis, the replication of an entity such as a molecule. If we look to chance chemical processes to assemble an autocatalytic molecule as the first step, then the high-pressure, high-temperature circumstance at depth would be favorable. There will be more and faster chemical interactions taking place than in cooler, lower-pressure situations, and hence the probability of setting up an extraordinary molecule will be greatest there. Furthermore, the number of different types of molecules that are stable is greatly enhanced by pressure. The deep, hot, high-pressure biosphere is the best place to look for the formation of such a vast array of different molecules, that the chance of forming an autocatalyst accidentally, perhaps only in long geologic times and in vast quantities of reactive materials, becomes a reasonable possibility. And a single autocatalytic molecule will dominate over all the rest in a small number of reproduction periods (like 100), as we have already seen above.

Life may have evolved as an exceedingly unlikely event of ordinary chemistry, but in circumstances where the number of chemical processes was so large that the unlikely became likely. We may think then of biospheres in other planetary bodies in which there are similar circumstances as those we see here. In such a picture the adaptation to surface life, and then to large and complex life forms, would take place only on a planet whose surface conditions fall into the narrow range of conditions that support complex life. But life at depth . . . that is another story.

Thomas Gold
Ithaca, New York
February 2001

Notes

CHAPTER 1

1. I first published the idea that hydrocarbons were not of biological origin in an op-ed piece, "Rethinking the origins of oil and gas," *Wall Street Journal,* June 8, 1977. The idea was more fully developed in my 1979 "Terrestrial sources of carbon and earthquake outgassing," *Journal of Petroleum Geology* 1(3): 3–19. See also Thomas Gold and Steven Soter, 1980. "The deep-earth gas hypothesis," *Scientific American* 242: 154–61. The idea is also the core of my 1987 book *Power from the Earth: Deep Earth Gas—Energy for the Future* (London: J.M. Dent & Sons).

2. The existence and naming of a "deep hot biosphere" were first proposed in Thomas Gold, 1992, "The deep, hot biosphere," *Proceedings of the National Academy of Sciences* 89: 6045–49.

3. P.N. Kropotkin reviews the history of the abiogenic theory of the origin of hydrocarbons in his 1997 "On the history of science: Professor N.A. Koudryavtsev and the development of the theory of origin of oil and gas," *Earth Sciences History* 16: 17–20.

4. My prediction of ten deep hot biospheres in the solar system appears in Thomas Gold, 1992, "The deep, hot biosphere," *Proceedings of the National Academy of Sciences* 89: 6045–49.

CHAPTER 2

1. A recent review article of the discovery and subsequent studies of the deep-ocean vent ecosystem is Daniel L. Distel, 1998, "Evolution of chemoau-

totrophic endosymbiosis in bivalves," *BioScience* 48(4): 277–86. See also the contributed chapters contained in David M. Karl, ed., 1995, *The Microbiology of Deep-Sea Hydrothermal Vents* (Boca Raton: CRC Press).

2. For a review of the publications that have reported methanotrophs in the deep-ocean vent communities, and as symbionts of the macrofauna, see Distell 1998, as cited in note 1.

3. Thomas D. Brock, 1978, *Thermophilic Microorganisms and Life at High Temperatures* (New York: Springer-Verlag).

4. C.K. Paull *et al.,* 1984, "Biological communities at the Florida Escarpment resemble hydrothermal vent taxa," *Science* 226: 965–67. Also, M.C. Kennicutt *et al.,* 1985, "Vent-type taxa in a hydrocarbon seep region on the Louisiana Slope," *Nature* 317: 351–52.

5. The discovery of the cave in Romania was reported in Serban M. Sarbu, Thomas C. Kane, and Brian K. Kinkle, 1996, "A chemoautotrophically based cave ecosystem," *Science* 272: 1953–55. See also media reports in the June 1996 issue of *Science News* (vol. 149, p. 405) and the January 1997 issue of *Discover* (p. 59).

6. The bacterial shrouds in the Mexican cave are reported in Charles Petit, 1998, "The walls are alive," *U.S. News and World Report,* February 9, pp. 59–60.

7. J. Cynan Ellis-Evans and David Wynn-Williams, 1996, "A great lake under the ice," *Nature* 381: 644–46.

8. NASA's interest in Lake Vostok is reported in Richard Stone, 1998, "Russian outpost readies for otherworldly quest," *Science* 279: 658–61.

9. Conditions for methane hydrate formation are presented in Ian R. MacDonald, 1997, "Bottom line for hydrocarbons," *Nature* 385: 389–90.

10. K.A. Kvenvolden and L.A. Barnard, 1982, "Hydrates of natural gas in continental margins," in J.S. Watkins and C.L. Drake, eds., *Studies in Continental Margin Geology,* AAPG Memoir 34, pp. 631–40.

11. A comparison of unoxidized carbon in methane hydrates versus other hydrocarbons is given in Carl Zimmer, 1997, "Their game is mud," *Discover,* May, pp. 28–30.

12. W. Steven Holbrook *et al.,* 1996, "Methane hydrate and free gas on the Blake Ridge from vertical seismic profiling," *Science* 273: 1840–43.

13. Accumulations of methane hydrates in permafrost and elsewhere are discussed in Yuri F. Makogan, 1981, *Hydrates of Natural Gas* (Tulsa: Penn Well Books).

14. The pink worms feeding on methane hydrates were reported in (anonymous), 1997, "Ice worms in the Gulf," *Science* 277: 769.

15. The threshold temperature for hyperthermophiles is defined in John A. Baross and James F. Holden, 1996, "Overview of hyperthermophiles and their heat-shock proteins," *Advances in Protein Chemistry* 48: 1–34.

16. John Postage, 1994, *The Outer Reaches of Life* (Cambridge, England: Cambridge University Press) p. 15.

17. That pressure may help maintain "the functional configuration of macro-molecules," is suggested in Baross and Holden (note 15).

18. Speculations on the upper temperature limit of life are presented in Baross and Holden 1996 (note 15).

19. Thomas Gold, 1992, "The deep, hot biosphere," *Proceedings of the National Academy of Sciences, USA* 89: 6045–49.

20. For a review of the debate about whether deep microbial life is indigenous or results from surface contamination, see John Parkes and James Maxwell, 1993, "Some like it hot (and oily)," *Nature* 365: 694–95.

21. S. L'Harldon *et al.*, 1995, "Hot subterranean biosphere in a continental oil reservoir," *Nature* 377: 223–24. The authors contend that because "the thermophilic isolates were repeatedly obtained from different wells and thrived in media similar to conditions in the wells" that "these isolates are members of a deep indigenous thermophilic community."

22. W.S. Fyfe, 1996, "The biosphere is going deep," *Science* 273: 448. An early and often-cited paper reporting hyperthermophilic life at a depth of 3 kilometers in Alaskan oil reservoirs is K.O. Stetter *et al.*, 1993, "Hyperthermophilic archaea are thriving in deep North Sea and Alaskan oil reservoirs," *Nature* 365: 743–45.

23. Richard Monastersky, 1997, "Signs of ancient life in deep, dark rock," *Science News* 152: 181.

24. U. Szevtzyk *et al.*, 1994, "Thermophilic, anaerobic bacteria isolated from a deep borehole in granite in Sweden," *Proceedings of the National Academy of Sciences, USA* 91: 1810–13.

25. The deep hot biosphere theory (see note 19) is cited, for example, in John Parkes and James Maxwell, 1993, "Some like it hot (and oily)," *Nature* 365: 694–95; in William J. Broad, 1993, "Strange new microbes hint at a vast subterranean world," *New York Times,* December 28, pp. C1, C14; Richard Monastersky, 1997, "Deep dwellers," *Science News* 151: 192–93; and in J.R. Delaney *et al.*, 1998, "The quantum event of oeanic crustal accretion: Impacts of diking at mid-ocean ridges," *Science* 281: 222–30.

26. Even when hydrocarbons (usually methane) are discovered in crystalline bedrock far from a sedimentary source, the hydrocarbons are presumed

to be of biological origin. Microbes feeding thereupon are thus judged to be dependent on buried organic products of the photosynthetic, surface biosphere. See, for example, Karsten Pedersen, 1996, "Investigations of subterranean bacteria in deep crystalline bedrock," *Canadian Journal of Microbiology* 42: 382–91. See also Delaney *et al.*, note 25.

27. Todd O. Stevens and James P. McKinley, 1995, "Lithoautotrophic microbial ecosystems in deep basalt aquifers," *Nature* 270: 450–54. See also the companion news piece by Jocelyn Kaiser, p. 377. And see James K. Fredrickson and Tullis C. Onstott, 1996, "Microbes deep inside the earth," *Scientific American,* October, pp. 68–73.

28. Stevens and McKinley (note 27) wrote that "High concentrations of dissolved methane have been observed locally in the Columbia River Basalt Group, and natural gas was commercially exploited early in this century, but the origin of the gas is uncertain."

29. Petra Rueter *et al.*, 1994, "Anaerobic oxidation of hydrocarbons in crude oil by new types of sulphate-reducing bacteria," *Nature* 372: 455–58.

30. The story of Carl Woese's successful effort to revise fundamentally the taxonomic classification of life is told in Virginia Morell, 1997, "Microbiology's scarred revolutionary," *Science* 276: 699–702.

31. C.J. Bult *et al.*, 1996, "Complete genome sequence of the methanogenic archaeon *Methanococcus jannaschii,*" *Science* 273: 1058–73.

CHAPTER 3

1. I have here adapted portions of an invited speech I delivered in 1988 at the IBM conference "Science and the Unexpected." The speech was later transcribed and published as "New ideas in science" in a 1989 issue of the *Journal of Scientific Exploration* 3(2):103–12. It was also published that same year as "The inertia of scientific thought," *Speculations in Science and Technology* 12:245–53.

2. My earliest writing on the deep-earth gas theory is my 1977 op. ed. essay "Rethinking the origins of oil and gas," *Wall Street Journal,* June 8. Later writings include Thomas Gold and Steven Soter, 1980, "The deep-earth gas hypothesis," *Scientific American* 242: 154–61; and Thomas Gold, 1985, "The origin of natural gas and petroleum and the prognosis for future supplies," *Annual Review of Energy* 10: 53–77.

3. Petroleum geologist and deep-gas entrepreneur Robert A. Hefner, III, has been very much interested in and supportive of the deep-earth gas theory from

the start. See his 1993 "New thinking about natural gas," in David G. Howell, ed., *The Future of Energy Gases,* USGS Professional Paper 1570.

4. Soon after developing my idea of a deep hot biosphere (then called "A hot, deep biosphere"), I submitted a paper with that title to *Nature.* This was June 1983. The paper was rejected. In May 1988 I tried again, this time with the title "Have we discovered a second domain of life on the earth?" I was of course referring to a second and independent biosphere in the deep earth. It was clear that rejection was inevitable for this paper, too, so I withdrew it and submitted it to a journal that is very reputable but not dominated by peer review. Any member of the National Academy of Sciences, as I was, could have a paper published, provided that concurrence could be obtained from two other members of the writer's choice. Thus in 1992, *Proceedings of the National Academy of Sciences, USA* (89: 6045–49) published my "Deep, hot biosphere" paper, upon the concurrence of Dr. Carl Woese and Dr. Gordon MacDonald.

5. E.M. Galimov, 1975, *Carbon Isotopes in Oil–Gas geology,* NTIS translation, pp. 335–36.

6. The formation of the earth and the primordial origin of hydrocarbons is covered in more detail in my chapter "Carbon—the element of life: What is its origin on earth?", Hermann Bondi and Miranda Weston-Smith, eds., 1993, *The Universe Unfolding* (Oxford, England: Clarendon Press).

7. This notion of a cool early earth is discussed in more detail in my 1985 paper "The origin of natural gas and petroleum, and the prognosis for future supplies," *Annual Review of Energy* 10: 53–77.

8. E.B. Chekaliuk, 1976, "The thermal stability of hydrocarbon systems in geothermodynamic conditions," in P. N. Kropotkin, ed., 1980, *Degasatsiia Zemli i Geotektonica* (Moscow: Nauka), pp. 267–72.

9. This discussion of thermodynamic stability of hydrocarbons at depth is drawn from my 1985 paper cited in note 6.

CHAPTER 4

1. This series of supporting claims for the abiogenic theory is more fully developed in my 1993 paper "The origin of methane in the crust of the earth," in David G. Howell, ed., *The Future of Energy Gases,* USGS Professional Paper 1570.

2. Koudryavtsev's rule is discussed in P.N. Kropotkin, 1997, "On the history of science: Professor N.A. Koudryavtsev and the development of the theory of origin of oil and gas," *Earth Sciences History* 16: 13–20.

3. Estimates of the possible volume of methane hydrates just below the surface in cold regions (and below the deep ocean) were made by Keith A. Kvenvolden, 1988, "Methane hydrate: A major reservoir of carbon in the shallow geosphere?" *Chemical Geology* 71: 41–51; and also by G.J. MacDonald, 1990, "Role of methane clathrates in past and future climates," *Climatic Change* 16: 247–81. See also Keith A. Kvenvolden, 1993, "A primer on gas hydrates," in David G. Howell, ed., *The Future of Energy Gases,* USGS Professional Paper 1570.

4. Refilling hydrocarbon reservoirs are documented in Robert F. Mahfoud and James N. Beck, 1995, "Why the Middle East fields may produce oil forever," *Offshore,* April, pp. 56–62. Documentation of refilling reservoirs along the U.S. Gulf Coast is provided in Jean K. Whelan, 1997, "The dynamic migration hypothesis," *Sea Technology,* September, pp. 10–18. See also Jean K. Whelan *et al.,* 1993, "Organic geochemical indicators of dynamic fluid flow processes in petroleum basins," *Advances in Organic Chemistry* 22: 587–615.

5. My discussion of carbonates is more fully developed in my 1993 USGS paper cited in note 1.

6. Among petroleum geologists, the view is widely held that in methane a deficiency of the heavy isotope by more than about 2 percent (20 per mil in the conventional units) characterizes this gas as being "unquestionably" of biogenic origin. This view is held despite many observational items that are in sharp conflict with it and despite the expectation that such fractionation would occur readily in migration of methane through tight rocks. How strongly this view is held by some, and how influential it has been, is shown, for example, in a paper by P.J. McCabe, D.L. Gautier, M.D. Lewan, and C. Turner (members of the U.S. Geological Survey) in "The future of energy gases," *USGS Circular* 1115, 1993. They conclude: "So far no economic accumulations of gas have been found that cannot be explained by the organic theory. Geochemical analysis from producing fields in the United States, for example, clearly shows that over 99 percent of the gas is of organic origin".

The question about the nature of the geochemical analysis involved that *clearly* showed this was never answered, despite repeated requests. The only types of analyses that might be involved would be the association with other gases and the carbon isotope ratio of the methane. The carbon isotope ratio is almost certainly the effect on which these authors hung this unsubstantiated statement, because quite likely over 99 percent of commercial methane shows a deficiency of carbon-13, greater than some arbitrary value that they deem to make the distinction between biogenic and abiogenic gases. Moreover, some commercial methane shows a much larger deficiency than can be explained by plant photosynthesis of any known vegetation (-75 per mil), so a fractionation along a migration path has to be invoked in any case.

The association with helium is not mentioned in their paper but would, in fact, rule out any biogenic origin. For example, the Texas Hugoton Field is a major natural gas–producing region, and the gas fields contain more helium than could have been produced by radioactivity in the sediments in their entire age, even if none had escaped. Only an ongoing supply of the gas mix from very deep levels can account for this.

The authors write in the same paper: "Clearly methane can be purely inorganic in nature and, in fact, *most scientists agree* that at least some of the methane on Earth is not of organic origin. Methane that emanates from mid-oceanic ridges, for example, contains what is *generally agreed* to be mantle-derived methane. But even though some inorganic methane is known to exist, *most scientists doubt* that commercial quantities of the gas ever escape the Earth's mantle because *carbon dioxide and water are the main fluids in the mantle.*" (The italics are mine.)

It is worth noting that the ocean-vent methane, which they agree is abiogenic, also has a substantial deficiency of carbon-13, ranging from -15.0 to -17.6 (in the usual notation), demonstrating that isotopic fractionation had occurred there also. In that case, why do we find mantle-derived diamonds and graphite? These are not produced from carbon dioxide and water. And why do we find much greater quantities of methane than of CO_2 in all deep holes that have been drilled?

7. The uniformity in isotopic ratios of carbonates of vastly different ages is discussed in M. Schidlowski, R. Eichmann, and C.E. Jung, 1975, "Pre-Cambrian sedimentary carbonates: Carbon and oxygen isotope geochemistry and implications for the terrestrial oxygen budget," *Precambrian Research* 2: 1–69.

8. Another point in favor of the abiogenic theory concerns the amount of free oxygen in the atmosphere. If all of the hydrocarbons within the earth's crust were attributable to the burial of surface life, then a very large surplus of free oxygen (a by-product of photosynthesis) would have been left behind in the atmosphere—more than 10 atmospheres of oxygen alone. The present atmosphere contains only about a fifth of 1 atmosphere of oxygen, and it seems doubtful that a surplus 50 times greater could have disappeared without leaving a clear record. If, on the other hand, the largest part of the deposits of unox-

idized carbon in the ground were deposited from unoxidized, carbon-bearing fluids coming up from below and not from materials taken down from the surface, then the discrepancy would disappear.

The strong association of helium with methane is presented in many papers of which I will give a selection here:

9. T. Gold and M. Held, 1987, "Helium-nitrogen-methane systematics in natural gases of Texas and Kansas," *Journal of Petroleum Geology* 10: 415–24.

10. J.A. Welhan and H. Craig, 1983, "Methane, hydrogen and helium in hydrothermal fluids at 21N on the East Pacific Rise," in P.A. Rona, ed., *Hydrothermal Processes at Seafloor Spreading Centers,* (Plenum Press), 391–409.

11. H. Craig gives a good overview of isotope separation process in various subsurface conditions and various gas mixes in his 1968 "Isotope separation by carrier diffusion," *Science* 159: 93–96 (January 5, 1968). There are several other fractionation processes for carbon isotopes discussed in the scientific literature that are beyond the scope of this book, but the processes based on diffusion speed seem to be the dominant ones.

12. V.F. Nikonov, 1969, "Relation of helium to petroleum hydrocarbons," *Dokl. Akad. Nauk SSSR.* Earth Sci. Sect. 188: 199–201.

13. V.F. Nikonov, 1973, "Formation of helium-bearing gases and trends in prospecting for them," *Internat. Geol. Rev.* 15: 534–41. Nikonov shows not only the high degree of association between natural hydrocarbon gases and oils but also that particular mixes of petroleum hydrocarbons are more enriched in helium than other types worldwide.

14. L.A. Pogorski and G.S. Quirt, 1978, "Helium emanometry in exploring for hydrocarbons: Part 1," *Proceedings of Symposium 1 on Unconventional Methods in Exploration for Petroleum and Natural Gas,* pp. 124–29.

15. A.A. Roberts, 1978, "Helium emanometry in exploring for hydrocarbons: Part II." *Proceedings of Symposium II on Unconventional Methods in Exploration for Petroleum and Natural Gas,* pp.136–49. The association of helium with hydrocarbons is shown also by the results that the helium seepage above oil and gas fields is so clear that helium surface measurements constitute a good method of locating hydrocarbon fields below. This demonstrates that helium is not just generally welling up in an area and just happens occasionally to be held in a trap that also holds hydrocarbons (as has sometimes been suggested) but rather that the surface presence of helium specifies the particular location of a hydrocarbon and helium field below.

16. W. Dyck and C.E. Dunn, 1986, "Helium and methane anomalies in domestic well waters in southwestern Saskatchewan, Canada, and their relationship to other dissolved constituents, oil and gas fields, and tectonic patterns," *Journal of Geophysical Research* 91: 12343–53.

The abundance ratio of the two stable isotopes of helium, helium-4, the common one, and the rare helium-3, has been measured in many locations (although the measurement requires very refined techniques). The results are of great significance for the discussion of the origin of hydrocarbons because a clear tendency exists for a slightly higher proportion of He-3 to be in helium that bears evidence of having come up from very deep levels. This effect has been given the explanation (almost certainly correctly) that small amounts of this gas come from a mix that was put into the forming earth and caught in the infalling solids. Solar and solar system gases contain the original isotopic mix supplied by the nuclear furnaces in stars, and this has a much higher concentration of He-3 than any helium on the surface of the earth, which was produced by the radioactive decay of uranium and thorium, making almost entirely He-4. The outgassing of helium will have removed the original mix from the outer layers, and hence we see elevated levels of helium-3 only from depths from which such outgassing has still not been complete.

While this allows us to identify sources of gas that contain an excess of He-3 as having come from some depth, it does not allow us to conclude that sources not showing a He-3 excess come from shallow levels. The mantle is not homogeneous and will have behaved quite differently in different regions. Helium outgassing is dependent on other gases having washed up through the pathways. A pathway that has long been flushed will have lost its primordial helium and just deliver the currently made He-4, while a pathway that has only been opened up in recent geological times will still show its primordial components. The heterogeneity of the mantle, the depth of origin of the gas, and the age of the pathway to the top will together define the isotopic ratio that will be observed. Hydrocarbon-rich areas are particularly prone to show elevated levels of He-3, and this makes a strong case that they have had a deep origin. Areas that do not have such an excess may have had a shallower origin, but they may also have had a deep origin from which they ascended on older, better swept pathways.

I am citing here a few of the papers from the large store of literature on the relations of He-3 to petroleum:

17. I.N. Tolstikhin, B.A. Mamyrin, E.A. Baskov, I.L. Kamenskii, G.A. Anufriev, and S.N. Surikov, 1975, "Helium isotopes, in gases from hot springs of the Kurile-Kamchatka volcanic region," in A.I. Tugarinov, ed., *Recent Contributions to Geochemistry and Analytical Chemistry,* (New York: John Wiley & Sons), 456–65.

18. I.N. Tolstikhin, 1975, "Helium isotopes in the Earth's interior and in the atmosphere: A degassing model of the Earth." *Earth Planet. Sci. Lett.* 26: 88–96.

19. H. Wakita and Y. Sano, 1983, "3He/4He ratios in CH_4-rich natural gases suggest magmatic origin," *Nature* 305: 792–94.

20. J.A. Welhan, W. Rison, R. Poreda, and H. Craig, 1983, "Geothermal gases of the Mud Volcano Area, Yellowstone National Park," *EOS* 64: 882.

21. H. Craig and J.E. Lupton, 1981, "Helium-3 and mantle volatiles in the ocean and the oceanic crust," *The Oceanic Lithosphere,* vol. 7. *The Sea* (New York: John Wiley & Sons), 391–428.

22. H. Craig, W.B. Clarke, and M.A. Beg, 1975, "Excess 3He in deep water on the East Pacific Rise," *Earth Planet. Sci. Lett.* 26: 125–32.

23. H. Craig, J.E. Lupton, J.A. Welhan, and R. Poreda, 1978, "Helium isotope ratios in Yellowstone and Lassen Park volcanic gases," *Geophys. Res. Lett.* 5: 897–900.

24. W. J. Jenkins, J. M. Edmond, and J. B. Corliss, 1978, "Excess 3He and 4He in Galapagos submarine hydrothermal waters," *Nature* 272: 156–58.

25. J.E. Lupton and H. Craig, 1975, "Excess 3He in oceanic basalts: Evidence for terrestrial primordial helium," *Earth Planet. Sci. Lett.* 26: 133–39.

26. J.E. Lupton, and H. Craig, 1981, "A major helium-3 source at 15S on the East Pacific Rise," *Science* 214: 13–18.

27. J.E. Lupton, G.P. Klinkhammer, W.R. Normark, R. Haymon, K.C. MacDonald, R.F. Weiss, and H. Craig, 1980, "Helium-3 and manganese at the 21N East Pacific Rise hydrothermal site," *Earth Planet. Sci.* 50: 115–27.

CHAPTER 5

1. This series of four supporting claims for the biogenic theory appears in my 1993 "The origin of methane in the crust of the earth," in David G. Howell, ed., *The Future of Energy Gases,* USGS Professional Paper 1570.

2. I developed the deep hot biosphere solution to the petroleum paradox over an extended period, beginning almost twenty years ago. In preparing these

notes, an assistant (Connie Barlow) and I culled through my files in an attempt to find written expressions of my transformation in thinking, when I was beginning to supplement the deep-earth gas theory with the deep hot biosphere theory but had not yet come to a full appreciation of the relationship. Two items caught our attention. In a transcript of an interview conducted by John Maddox for the BBC and broadcast in June 1978, as part of the "Scientifically Speaking" radio program, it is clear that I had not yet begun to entertain the idea that microbes might live at depth—or at least I wasn't willing to voice this hypothesis. I said, "In recent years, one has found a lot more gas, deeper than any oil and very methane-clean with no other hydrocarbons. I tend to think that mostly it is primeval material, but of course the fact that there is undoubtedly biogenic hydrocarbon in the ground makes it very hard to distinguish the two."

Five years later I was ready to take the discussion much further. In an interview published in the March 1983 issue of *Montana Oil Journal*, I said this: "The genuinely biological content of most oils is only a small fraction and [is] by no means difficult to account for. When an oil is in fossiliferous sediments, it will certainly leach out all the oil-soluble biological material. Also, oil is a very desirable substance for various forms of microbiology, and we see clearly that where the temperature of the oil is low enough for this to flourish, the biological markers are present."

Two months later, in June 1983, I submitted my paper, "A hot, deep biosphere," to *Nature*. Refer to note 4 in Chapter 3 for the history of my attempt to get this idea published.

3. Guy Ourisson, Pierre Albrecht, and Michel Rohmer, August 1984, "The microbial origin of fossil fuels," *Scientific American* 251(2): 44–51.

4. My reply to the Ourrison *et al.* paper was published in November 1984, *Scientific American* 251(5): 6.

5. In 1984, at the time of the Ourisson paper, no distinction had yet been made between bacteria and archaea.

6. Robert Robinson, 1963, "Duplex origin of petroleum," *Nature* 199: 113–14.

7. My estimate of the microbial biomass at depth was published in Thomas Gold, 1992, "The deep, hot biosphere," *Proceedings of the National Academy of Sciences* 89: 6045–49.

8, While this book was in the page proof stage, an important paper was published that qualitatively corroborates my projections of a very large biomass contained within the deep hot biosphere: J.R. Delaney *et al.,* 1998, "The quantum event of oceanic crustal accretion: Impacts of diking at mid-ocean ridges,"

Science 281: 222–30. The authors describe a previously unknown phenomenon of sudden and massive release of hydrothermal fluids on the ocean floor. Because these fluids contained "massive effusions of microbial products," the authors inferred that the biological activity must have occurred *before* the fluid ejection, in "warmer subseafloor habitats." Their conclusion: "The zone in the crust occupied by thermophiles may be extensive." They also stated, "Massive and sustained output of microbial products associated with diking lend support to recent postulates of a significant deep hot biosphere within the Earth." My 1992 paper is among the references cited with that statement.

Because of my contention (in Chapter 8) that upwelling plumes of methane are the cause of many earthquakes, the observation by Delaney *et al.* that "earthquake swarms" seem to accompany the release of the hydrothermal fluids is of great interest to me.

9. Ourisson *et al.* wrote, "Because of the difference in age and because the organic compounds in coal and petroleum were thought to come from different sources, the correspondence of peaks in the C_{27} to C_{32} region was unexpected." Guy Ourisson, Pierre Albrecht, and Michel Rohmer, 1984, "The microbial origin of fossil fuels," *Scientific American* 251 (2): 44–51.

10. K.R. Pedersen and J. Lam, 1970, "Precambrian organic compounds from the Ketilidian of south-west Greenland," *Gronlands Geologiske Unders. Bull.,* No. 82.

11. G. Henderson, 1964, "Oil and gas prospects in the Cretaceous–Tertiary basin of west Greenland," *Geol. Survey Greenland Rept.,* No. 22.

12. C.H. Hitchcock, 1865, "The Albert Coal, or Albertite, of New Brunswick," *Amer. J. Sci, 2nd Ser.* 39: 267–73.

13. These and many more examples of anomalies in coal deposits are discussed in Chapter 9 of my 1987 book, *Power from the Earth* (London: J.M. Dent).

14. H.R. Wanless, 1952, "Studies of field relations of coal beds," in *Second Conference on the Origin and Constitution of Coal,* Nova Scotia Department of Mines, pp. 148–80.

CHAPTER 6

1. My presentation was published in Sweden in *Svenska Dagbladet* (October 17, 1983) under the title "Deep natural gas in Sweden?" (file no. 235)

2. I published a more detailed account of the Siljan results in the January 14, 1991 issue of *Oil and Gas Journal,* pp. 76–78. See also my 1988 "The deep earth gas theory with respect to the results from the Gravberg-1 well," in A.

Bodén and K.G. Eriksson, *Deep Drilling in Crystalline Bedrock* (New York: Springer-Verlag) pp. 18–27.

3. I wrote a summary of the findings on the magnetite sludge in 1991, "Accumulations of fine-grained magnetite in a deep borehole in Sweden," unpublished. (file no. 275)

4. The two laboratory analyses were performed by J.M. Knudsen *et al.* at Orsted Laboratory at the University of Copenhagen and by R. Reynolds *et al.* at the USGS in Denver. Both are contained in my file no. 652.

5. An account of a third laboratory analysis, including neutron activation analysis, performed by C. Orth at the Los Alamos National Laboratory, was communicated to me in a private report; file no. 652.

6. A summary of the Philp findings are contained in my green file no. 652.

7. U. Szewzyk *et al.,* 1994, "Thermophilic, anaerobic bacteria isolated from a deep borehole in granite in Sweden," *Proceedings of the National Academy of Sciences* (USA) 91: 1810–13.

8. During the time of Siljan drilling, announcement was made of the biological production of ultrafine-grain magnetite by anaerobic microbes in Derek R. Lovley, John F. Stolz, Gordon L. Nord, and Elizabeth J.P. Phillips, 1987, "Anaerobic production of magnetite by a dissimilatory iron-reducing microorganism," *Nature* 330: 252–54. See also N.H.C. Sparks *et al.,* 1990, "Structure and morphology of magnetite anaerobically produced by a marine magnetotactic bacterium and a dissimilatory iron-reducing bacterium," *Earth and Planetary Science Letters* 98: 14–22.

9. W.E. Henry, 1989, "Magnetic detection of hydrocarbon microseepage in a frontier exploration region," *Bulletin of the Association of Petroleum Geochemical Explorationists* 5: 18–29.

10. P.N. Kropotkin, 1997, "On the history of science: Professor N.A. Koudryavtsev (1893–1971) and the development of the theory of origin of oil and gas," *Earth Sciences History* 16(1):17–20.

CHAPTER 7

1. D.G. Pearson, G.R. Davies, P.H. Nixon, and H.J. Milledge, 1998, "Graphitized diamond from a peridotite massif in Morocco and implications for anomalous diamond occurrences," *Nature* 338: 60–62.

2. C.E. Melton and A.A. Giardini, 1974, "The composition and significance of gas released from natural diamonds from Africa and Brazil," *American Mineralogist* 59: 775–82.

3. Pierre Cartigny, Jeffrey W. Harris, and Marc Javoy, 1998, "Eclogitic diamond formation at Jwaneng: No room for a recycled component." Science 280: 1421–23.

4. Konrad B. Krauskopf, 1982, *Introduction to Geochemistry* (London: McGraw-Hill) p. 395.

5. P.N. Kropotkin, 1997, "On the history of science: Professor N.A. Koudryavtsev and the development of the theory of the origin of oil and gas," *Earth Sciences History* 16: 13–20.

6. An important example of a scientific paper that reports an association between gold and hydrocarbons (and yet assumes a hydrothermal cause for the metal deposit) is A.C. Barnicoat *et al.*, 1997, "Hydrothermal gold mineralization in the Witwatersrand Basin," *Nature* 386: 820–24.

7. Tables of biomineralization can be found in Barry S.C. Leadbeater and Robert Riding, eds., 1986, *Biomineralization in Lower Plants and Animals* (Oxford, England: Clarendon) p. 4; and also in Lynn Margulis and Dorion Sagan, 1995, *What Is Life?* (New York: Simon & Schuster).

CHAPTER 8

1. Mud volcanoes are discussed in numerous Russian publications. In the English-language scientific literature an interesting, illustrated article is Martin Hovland, Andrew Hill, and David Stokes, 1997, "The structure and geomorphology of the Dashgill mud volcano, Azerbaijan," *Geomorphology* 21:1–15. Mud volcanoes are also discussed in my 1987 book, *Power from the Earth.*

2. The most comprehensive account of pockmarks and other ocean floor features that I interpret as caused by the release of gases, mostly methane, is contained in M. Hoveland and A.G. Judd, 1988, *Seabed Pockmarks and Seepages: Impact on Geology, Biology, and the Marine Environment* (London: Graham and Trotman). The largest known pockmark field and the implications of its methane release are discussed in S. Lammers, E. Suess, and M. Hovland, 1995, "A large methane plume east of Bear Island (Barents Sea): Implications for the marine cycle," *Geol. Rundsch* 84:59–66.

3. Martin Hovland *et al.*, 1994, "Fault-associated seabed mounds (carbonate knolls?) off western Ireland and northwest Australia," *Marine and Petroleum Geology* 11(2):232–46.

4. I thank Steven Soter for alerting me to the temporal correlation between the decline in the gas theory of earthquakes and the invention of seismographs.

5. Over the course of many years, Steven Soter has researched earthquake data and has collaborated in papers and articles on deep gas as a cause of earth-

quakes. These writings include Thomas Gold and Steven Soter, 1980, "The deep-earth gas hypothesis," *Scientific American* 242: 154–61; and Thomas Gold and Steven Soter, 1984/85, "Fluid ascent through the solid lithosphere and its relations to earthquakes," *Pageoph* 122: 492–530. See also the chapter on earthquakes in my 1987 book, *Power from the Earth* (London: J.M. Dent).

6. Isaac Newton, 1730, *Optics,* 4th ed., q. 31, pp. 354–55.

7. John Michell, 1761, "Conjectures concerning the cause, and observations upon the phenomena, of earthquakes," *Philosophical Transactions of the Royal Society* 51: 566–634.

8. Liao-ling Meteorological Station, 1977, "The extraordinary phenomena in weather observed before the February 1975 Hai-cheng earthquake," *Acta Geophys. Sinica* 20: 270–75.

9. Yuji Sano and Hiroshi Wakita, 1987, "Helium isotope evidence for magmatic gases in Lake Nyos, Cameroon," *Geophysical Research Letters* 14: 1039–41.

10. Unfortunately, I have only an incomplete citation for the Tangshan earthquake and could not retrieve the full citation. The quotation was drawn from a report by J. Li, titled "Earthquake: A harvest of agony," which appeared in an October 1980 issue of the *Los Angeles Times*.

11. I understand that earth mounds are present in other regions of North America, too. Interpretations tend to identify the cause of their formation as mysterious, though perhaps attributable to earthworks deliberately constructed by paleo Indians.

12. For a review of strain mechanisms hypothesized for earthquakes and the problems of using such mechanisms to explain deep earthquakes, see Harry W. Green II and Heidi Houston, 1995, "The mechanics of deep earthquakes," *Annual Reviews of Earth and Planetary Sciences* 23: 169–213. The authors share my view that pressurized fluids in pore spaces can account for earthquakes at depths in which strain cannot play a role, but they posit water rather than hydrocarbons as the permeating fluid. As explained in Chapter 3, I also disagree with the authors' contention that "it is extremely unlikely that a fluid-filled porosity . . . could be maintained to great depth."

CHAPTER 9

1. Infrared phototaxis (direction finding by radiation) at hot ocean vents is suggested as a precursor to photosynthesis in Euan G. Nisbet, 1995, "Archaean ecology: A review of evidence for the early development of bacterial biomes, and speculations on the development of a global-scale biosphere," in M.P.

Coward and A.C. Ries, eds., *Early Precambrian Processes,* Geological Society of London Special Publication No. 95, pp. 27–51. See also E.G. Nisbet and C.M.R. Fowler, 1996, "Some liked it hot," *Nature* 382: 404–5.

2. The calculation assumes that each monkey is given just one chance to push, at random, the requisite number of keys (39) corresponding to the first line in a Shakespeare sonnet.

3. Robert V. Miller, 1998, "Bacterial gene swapping in nature," *Scientific American,* January, pp. 67–71.

4. Vilmos Csányi proposed that symbiotic mergers of communities of microorganisms might best account for fundamental differences in cell types of the animal body in his 1989 *Evolutionary Systems and Society* (Durham, NC: Duke University Press).

5. Linus Pauling, 1959, *General Chemistry,* 2nd ed. (San Francisco: W.H. Freeman), p. 599.

6. Support for the gene-swapping view of early life is reported in Elizabeth Pernise, 1998, "Genome data shake tree of life," *Science* 280: 672–74.

CHAPTER 10

1. Jon S. Nelson and E.C. Simmons, 1995, "Diffusion of methane and ethane through the reservoir cap rock: Implications for the timing and duration of catagenesis," *American Association of Petroleum Geologists Bulletin,* July 1995, 79(7):11064–74.

2. David M. Karl, 1995, "Ecology of free-living, hydrothermal vent microbial communities," in David M. Karl, ed., *The Microbiology of Deep-Sea Hydrothermal Vents* (Boca Raton, FL: CRC Press), p. 109.

3. For a review of the publications that have reported methanotrophs in the deep-ocean vent communities and as symbionts of the macrofauna, see Daniel L. Distel, 1998, "Evolution of chemoautotrophic endosymbiosis in bivalves," *BioScience* 48(4): 277–86.

4. John Postgate, 1994, *The Outer Reaches of Life* (Cambridge, England: Cambridge University Press), p. 29.

5. Climate model calculations for Venus are reported in James F. Kasting, 1988, "Runaway and moist greenhouse atmospheres and the evolution of Earth and Venus," *Icarus* 74: 472–94. See also his 1997 "Habitable zones around low mass stars and the search for extraterrestrial life," *Origins of Life and Evolution of the Biosphere* 27: 291–307.

6. The time for onset of a runaway greenhouse on the earth has been estimated by James Lovelock and Michael Whitfield in their 1982 "Life span of the biosphere," *Nature* 296: 561–63. A more recent estimate is given in Ken Caldeira and James F. Kasting, 1992, "The lifespan of the biosphere revisited," *Nature* 360: 721–23.

7. Baerbel K. Lucchitta, Duwayne M. Anderson, and Hitoshi Shoji presented a strong case for Mars glaciation in their 1981 "Did ice streams carve Martian outflow channels?" *Nature* 290: 759–63. See also Hugh H. Kieffer *et al.,* eds, *Mars* (Tucson: University of Arizona Press) pp. 498–51. The glacial theory is applied to the most recent Mars photos in Jeffrey Winters, 1998, "A survey of ancient Mars," *Discover,* July, pp. 113–17.

8. The gigantic magnitude of Olympus Mons on Mars creates a difficulty in accounting for it as a lava volcano. Because lava is nearly the same density as the solid rock, it is difficult to imagine how an adequate pressure could be available to push lava to the height of the landform in its later stages. Possibly it should be thought of as a giant pingo or a giant mud volcano.

9. D.S. McKay *et al.,* 1996, "Search for Past Life on Mars: Possible Relic Biogenic Activity in Martian Meteorite," ALH 84001. *Science* 273(5277):1 924–30, 16 August 1996

10. Thomas Gold, 1992, "The deep, hot biosphere," *Proceedings of the National Academy of Sciences, USA* 89: 6045–49.

Acknowledgments

T he assistance I received from Connie Barlow in writing about this diverse and multi-
faceted subject is greatly appreciated. Her
suggestions were responsible for ordering my views and ideas into
sequences that could be understood more readily, and she added several significant explanations and found essential references in my
extensive files.

I am greatly indebted to Mr. William Frucht, past editor of Copernicus, who developed great interest in the subject of massive subsurface life
and the numerous consequences this would have in various branches of
the earth sciences. He persuaded me to write the present book.

Luckily the present chief editor of Copernicus, Mr. Jonathan Cobb,
took an equally strong interest in the subject, and his editorial pen had
a major impact on the brevity and clarity of the explanations. I would
also like to thank Connie Day for her many fine copyediting suggestions.

Ralph E. Gomory, the president of the Alfred P. Sloan Foundation,
was also enthusiastic about the subject, as was Dr. Jesse Ausubel, a
member of the staff, and I greatly appreciate that numerous expenses
incurred with the collection of the evidence were covered by that foundation.

I discovered some time ago that I was not the originator of the theory of a deep origin of petroleum; Russian or ex–Soviet Union scientists

had concerned themselves for more than a century with these subjects and had obtained truly vast amounts of information in support of them. In particular, Peter Kropotkin (now deceased), a distinguished geologist in the Geological Institute of the U.S.S.R. Academy of Sciences, was my main source of information. I was gratified to have been asked by the U.S.S.R. Academy to contribute a chapter to a book in memory of Mendeleyev, who was one of the first to point to such a theory.

The past assistance of my friend and colleague Dr. Steven Soter constituted major contributions in many areas, especially in relation to earthquakes, on which he had collected eyewitness accounts over centuries from many lands and, by showing me the breadth of viewpoints, greatly encouraged me to think the problems out afresh and find the inadequacies of many commonly held explanations.

In considering the presence of huge amounts of subsurface microbiology, necessary to account for biological molecules in all petroleum, I was influenced by the remarkable work of Professor Ourisson and his collaborators in Strasbourg, France, who had demonstrated that very large amounts of bacterial debris existed in the rocks together with oil and gas, although they placed a different explanation on these finds.

I also wish to thank my wife for the great patience she showed when I secluded myself in my study for days on end, and when my time-table sometimes interfered with the orderly conduct of the household.

Index